International Conference on Mechanical Engineering and Modern Technologies (MEMT2020)

AA007573

IOP Conference Series: Materials Science and Engineering
Volume 1118

Tomsk, Russia
26 - 30 October 2020

ISBN: 978-1-7138-3404-5
ISSN: 1757-8981

Printed from e-media with permission by:

Curran Associates, Inc.
57 Morehouse Lane
Red Hook, NY 12571

Some format issues inherent in the e-media version may also appear in this print version.

This work is licensed under a Creative Commons Attribution 3.0 International Licence.
Licence details: http://creativecommons.org/licenses/by/3.0/.

No changes have been made to the content of these proceedings. There may be changes to pagination and minor adjustments for aesthetics.

Printed with permission by Curran Associates, Inc. (2021)

For permission requests, please contact the Institute of Physics at the address below.

Institute of Physics
Dirac House, Temple Back
Bristol BS1 6BE UK

Phone: 44 1 17 929 7481
Fax: 44 1 17 920 0979

techtracking@iop.org

Additional copies of this publication are available from:

Curran Associates, Inc.
57 Morehouse Lane
Red Hook, NY 12571 USA
Phone: 845-758-0400
Fax: 845-758-2633
Email: curran@proceedings.com
Web: www.proceedings.com

TABLE OF CONTENTS

SELF-CONTAINED DRILLING ASSEMBLY 1
L Yu Belozertseva

DETERMINATION OF ROCK FRACTURE PARAMETERS IN DIRECTIONAL PERCUSSION DRILLING WITH DRILLHOLE PATH ADJUSTABILITY 7
E M Chernienkov

INVESTIGATION OF THE CRITERIA FOR EVALUATING ELECTRODE METAL TRANSFER IN SHORT CIRCUIT GAS-SHIELDED ARC WELDING 14
S V Bolotov

ION-PLASMA TREATED PARTS QUALITY IMPROVEMENT ANALYSIS BASED ON THE RELIABILITY THEORY CRITERIA 20
V I Bogdanovich, M G Giorbelidze

ION-PLASMA COATINGS PERFORMANCE PROPERTIES IMPROVEMENT OBTAINED BY ARC DEPOSITION 25
V I Bogdanovich, M G Giorbelidze

ANALYSIS OF CONTACT STRENGTH OF SPHERICAL ROLLER TRANSMISSION WITH DOUBLE-ROW PINION 30
M E Lustenkov, A N Moiseenko

EXPERIMENTAL ESTIMATION OF EFFICIENCY AND KINEMATIC ACCURACY OF A SPHERICAL ROLLER TRANSMISSION 36
E S Lustenkova

SIMULATION OF SURFACE WORK-HARDENING PROCESS DURING CONTROLLED INERTIAL IMPACT TREATMENT OF PARTS MADE OF HIGH STRENGTH CAST IRON 42
V M Pashkevich, V V Afanevich

EFFECT OF GLOW DISCHARGE TREATMENT ON THE DEPTH AND DEGREE OF STRUCTURAL PHASE MODIFICATION OF TOOL STEEL PRODUCTS 48
V M Shemenkov, M A Rabyko

SURFACE TREATMENT BY COMBINED PROCESSES OF CUTTING AND MAGNETODYNAMIC ROLLING 55
D M Svirepa, A S Kruchynina

EFFECT OF HOT ISOSTATIC PRESSING ON THE QUALITY OF METAL PARTS OBTAINED BY SELECTIVE LASER MELTING 60
A V Zhdanov, A A Voznesenskaya, A S Raznoschikov

STABILIZATION OF ROOT PARAMETERS FOR SHIELDED ARC WELDING 65
P P Krasikov, A V Savinov, O A Polesskiy, A A Chudin, L S Krasikova, I V Kozlov, D S Borisov, V V Filippov

COLD DEFORMATION OF NON-CIRCULAR DRILL PIPES 70
S V Parshin, A I Dronov

ILABIT ROBOTICA: A PHOTO-REALISTIC EXTENSION FOR ROBOTICS TOOLBOX 74
Ivan Kholodilin, Nikita Savosteenko, Yuri Abramov, Luis A Lago

ESTIMATION OF ELECTRIC LOADS DYNAMICS IN THE GAS FIELDS OF WESTERN SIBERIA TAKING INTO ACCOUNT THE LIFE CYCLE OF THE FIELDS 80
I M Bogachkov, R N Khamitov, V P Frayshteter

STUDYING INTERACTION OF MULTI-MOTOR FORMING SYSTEM FOR DOUBLE CURVATURE ITEMS 90
N E Deryuzhkova, V A Solovyev, V V Teterin

COMPARISON OF CONTROL METHODS FOR INVERTED 2-DEGREE OF FREEDOM PENDULUM MOUNTED ON THE CART 97
O Y Sumenkov, A S Belyaev

THE DESIGN OF A SUPPORT CHAIR FOR MORE EFFECTIVE PATIENT REHABILITATION 103
G Mamontov, V Seryakov, Y Khmelevsky, L Tkacheva

QUALITY RATIOS FOR PUBLIC TRANSPORT SERVICES 109
A N Novikov, S V Eremin, A G Shevtsova, D O Lomakin, A V Kulev

PATH-BUILDING AND LOCALIZATION SYSTEM FOR MOBILE ROBOTIC PLATFORMS BASED ON RASPBERRY PI 113
E A Ivanov, A S Belyaev, O A Brylev

EVALUATION OF THE POSSIBILITY OF USING MULTILAYER NANOSTRUCTURED COATINGS TO IMPROVE THE OUTPUT CHARACTERISTICS OF ROLLER SCREW MECHANISMS 119
A V Zhdanov, E A Novikova, N S Dovbysh, A N Mitrofanov

A DEVICE FOR AUTOMATIC FEEDING OF COATED ELECTRODES INTO A WELD POOL 124
A S Gordynets, A S Nepomnyashiy

RESULTS OF MODELING THE OPERATION OF THE HYDRAULIC PUMP IN THE MATLAB-SIMULINK APPLICATION AND TESTING ON THE DIAGNOSTIC BENCH 128
Yu Vlasov, M Meshcheriakov, T Alushkin

THERMAL CONDUCTIVITY AND HEAT CAPACITY OF NANOFLUID BASED ON WATER MODIFIED BY HYBRID MATERIAL OF COMPOSITION DETONATION NANODIAMONDS-CARBON NANOTUBES 132
A A Vozniakovskii, A P Voznyakovskii, S V Kidalov, E K Kalashnikova

STUDY OF THE MAGNETIC PROPERTY OF SUPERMAGNETIC FLUID 137
O K Kuvandikov, Sh J Kuvandikov, M K Salakhitdinova, Kh A Kayumov

DETERMINATION OF OPTIMAL FREQUENCY TO REPLACE CUTTING ELEMENTS IN CHAIN TRENCHERS 141
S M Kravchenko, V A Slepchenko, I V Slepchenko

EFFICIENT TECHNOLOGIES FOR PRODUCING CAST IRON BILLETS AND PRODUCTS WITH SPECIFIED PROPERTIES AND MICROSTRUCTURE 146
D A Lubyanoi, A V Markidonov, E Yu Pudov, V N Shakhmanov, E G Kuzin, O S Semenova

FAILURES OF SAFETY SERVICES IN CASE OF EMERGENCY FALL OF CARGO-PASSENGER ELEVATOR'S CABIN EXPLOITED IN INDUSTRIAL AND CIVIL PREMISES 151
M Popov, A Negodin

SIMULATION ON ELEVATOR EMERGENCY BRAKE 155
M Popov, A Negodin

THE STUDY OF THE POWER SUPPLY SYSTEM OF THE EXISTING GAS FIELDS IN WESTERN SIBERIA USING THE THEORY OF EXPERIMENTAL PLANNING 159
I M Bogachkov

DEVELOPING THE CALCULATION METHODS OF EFFECTIVE VALUES OF CURRENT AND VOLTAGE FOR NONSINUSOIDAL TRANSIENT MODES IN ELECTRIC POWER SYSTEMS BASED WAVELET TRANSFORM 166
O V Arkhipova, N N Dolgikh, V Z Kovalev, D S Osipov, A O Paramzin, V A Tkachenko

THE POSSIBILITY TO USE BANDSAW DYNAMIC TENSIONING IN BANDSAW MACHINES 172
V K Shilko, V A Slepchenko, I V Slepchenko

INFLUENCE OF SOME PARAMETERS OF MAGNETIZATION AND CUTTING MODES ON THE TOOL LIFE WHEN MACHINING CONSTRUCTION MATERIALS 177
S M Khasanov, A A Yakubov

FLEXURAL PROPERTIES OF COPPER-FILLED CARBON FIBER REINFORCED POLYMERS 183
A V Eremin, A V Burkov

THERMOGRAVIMETRIC ANALYSIS OF EPOXY-BASED CARBON FIBER REINFORCED POLYMERS MODIFIED BY CARBON FILLERS 188
M V Burkov, A V Eremin

Author Index

Preface

The International Conference on Mechanical Engineering and Modern Technologies 2020 (MEMT 2020)

26-30 October 2020
Tomsk, Russia

The International Conference on Mechanical Engineering and Modern Technologies 2020 (MEMT 2020) was held at Tomsk Polytechnic University, Tomsk. Russia on October 26–30, 2020. MEMT 2020 is a scientific forum where researchers, students and industry professionals exchange ideas, share experience and discuss technical issues in the field of mechanical engineering, material sciences, modelling

The conference was organized partially by personal attendance of participants as well as virtually using Internet communication technologies (Zoom). The conference attendance was organized in compliance with all sanitary standards and safety requirements. The majority of participants attended the Conference personally. It is worth noting that the option of virtual participation allowed expanding significantly the geography and increasing the number of participants.

The conference program included as well Online Tutorial Workshop "Advanced Production Technologies", Round Table "Implementation of Technical Disciplines Using Remote Educational Resources" and the workshop for young scientists “Modern Technologies and Equipment”. The representatives of such companies as Ariel Corporation, The sessions covered the following topics:

- Machine and mechanisms reliability;
- Electrical engineering and mechatronics;
- New manufacturing technology;
- Materials science and nanotechnology in mechanic engineering;
- Automation and modeling of engineering processes;
- Industrial safety;
- Design in mechanic engineering.

We appreciate contributions from all participants of MEMT 2020. We also want to thank sponsors for the financial support and Program Committee members for their efforts in organization and holding of the MEMT 2020 Conference.

Editors:
- Dr. Evgueny N. Pashkov, epashkov@tpu.ru, Tomsk Polytechnic University
- Dr. S.N. Sorokova, s_sorokova@tpu.ru, Tomsk Polytechnic University
- Dr. S.N. Kladiev, kladiev@tpu.ru, Tomsk Polytechnic University
- Dr. Viktor N. Kozlov, kovn@tpu.ru, Tomsk Polytechnic University
- Dr. Egor A. Efremenkov, egorefr@tpu.ru, Tomsk Polytechnic University
- Prof. Maria S. Kukhta, kuhta@tpu.ru, Tomsk Polytechnic University
- Dr. E.A. Kulinich, kulinich@tpu.ru, Tomsk Polytechnic University
- Anna Bogdan, coalise@tpu.ru, Tomsk Polytechnic University

Content from this work may be used under the terms of the Creative Commons Attribution 3.0 licence. Any further distribution of this work must maintain attribution to the author(s) and the title of the work, journal citation and DOI.
Published under licence by IOP Publishing Ltd

IOP Conf. Series: Materials Science and Engineering 1118 (2021) 011001 doi:10.1088/1757-899X/1118/1/011001

ORGANIZING COMMITTEE

Organizing Committee Chair: Dr. Evgueny N. Pashkov, epashkov@tpu.ru, Tomsk Polytechnic University.

Members:

Dr. S.N. Sorokova, associate professor, School of Advanced Manufacturing Technologies, Tomsk Polytechnic University;
Dr. E.A. Efremenkov, associate professor, School of Advanced Manufacturing Technologies, Tomsk Polytechnic University;
Prof. A.I. Sechin, professor, School of Computer Science & Robotics, Tomsk Polytechnic University;
Dr. V.N. Kozlov, associate professor, School of Advanced Manufacturing Technologies, Tomsk Polytechnic University;
Dr. V.N. Deryusheva, associate professor, School of Advanced Manufacturing Technologies, Tomsk Polytechnic University;
Dr. E.A. Kulinich, Academic Secretary, School of Nuclear Science & Engineering, Tomsk Polytechnic University;
Dr. S.N. Kladiev, associate professor, School of Energy & Power Engineering, Tomsk Polytechnic University;
Dr. P.V. Isaenko, associate professor, Tomsk State University of Architecture and Building
A.M. Bogdan, expert, School of Computer Science & Robotics, Tomsk Polytechnic University;
M.S. Cheremiskina, assistance lecturer, School of Core Engineering Education, Tomsk Polytechnic University

PROGRAMME COMMITTEE

Programme Chair: prof. Sergey V. Panin, Deputy Director for Research of ISPMS SB RAS, Doctor of Technical Sciences, Professor, Tomsk, Russia.

Members:

Prof. M.V. Kurlenya, Academician of the Russian Academy of Sciences, Professor of the IGD SB RAS Novosibirsk, Russia;
Prof. V. Popov, Professor at the Technical University of Berlin (Germany);
Prof. M.E. Lustenkov, Rector of the Belarusian-Russian University (Belarus);
Prof. V.I. Klishin, Corresponding Member of the Russian Academy of Sciences, Director of the Institute of Coal SB RAS, Kemerovo, Russia;
Dr. D.A. Chinakhov, Director of the Yurga Technological Institute, Yurga, Russia;
Prof. A.B. Efremenkov, Vice-Rector for Research and Innovation, NovSU, Doctor of Technical Sciences, Veliky Novgorod, Russia;
Prof. Yu.A. Vlasov, Dean of the Faculty of Mechanics and Technology of TSASU, Doctor of Technical Sciences, Professor, Tomsk, Russia;
Prof. M.V. Dudkin, Dean of the Faculty of Mechanics and Technology of EKSTU, Doctor of Technical Sciences, Professor, Ust-Kamenogorsk, Kazakhstan;
Prof. Joel Smeturst, Professor at the University of Southampton, Southampton, UK;
Legun I.A., Regional Manager for Russia and CIS Ariel (Ariel International Corporation), Moscow, Russia;
Pavel Stibor, Executive Director of Domist company, Czech Republic
Prof. Saidaminov Isohon, Professor of the Tajik University of Technology, Tajikistan;
Prof. Nguyen Huu Locke, Ho Chi Minh University of Technology, Vietnam;
Prof. Tomilin A.K., Doctor of Physical and Mathematical Sciences, Tomsk Polytechnic University, Russia;
Prof. Knyazeva A.G., Doctor of Physical and Mathematical Sciences, Tomsk Polytechnic University, Russia;
Prof. Kukhta M.S., professor of the Department of Material science, Tomsk Polytechnic University, Russia

IOP Conf. Series: Materials Science and Engineering 1118 (2021) 011002 doi:10.1088/1757-899X/1118/1/011002

Peer review declaration

All papers published in this volume of *IOP Conference Series: Materials Science and Engineering* have been peer reviewed through processes administered by the Editors. Reviews were conducted by expert referees to the professional and scientific standards expected of a proceedings journal published by IOP Publishing.

- **Type of peer review: Single-blind (**We had a two-level review system: first, the members of Program Committee reviewed papers by the relevance to the conference topics and paper format requirements; the second review process involved independent reviewers who evaluated papers according to the following criteria as the relevance of study, scientific quality, references adequacy, paper structure, etc., then papers were also checked by English-speaking persons)

- **Conference submission management system:** the papers were submitted via conference website

- **Number of submissions received:** 81

- **Number of submissions sent for review:** 76

- **Number of submissions accepted:** 37

- **Acceptance Rate (Number of Submissions Accepted / Number of Submissions Received X 100):** 48%

- **Average number of reviews per paper:** 2

- **Total number of reviewers involved:** 6

- **Any additional info on review process:**

  Firstly, Members of Program Committee (Editors) reviewed all submitted papers by the relevance to the conference topics and paper standard requirements. They also check submissions for plagiarism and duplicates. The manuscripts submitted should meet the followings criteria:
  - Papers should be written in English;
  - Papers should be formatted according to the journal's template;
  - Scientific merit and data quality;
  - Good quality of graphic images;
  - Structure of the article should be appropriate and clear.

  Secondly, papers were reviewed by experts in relevant fields of science. They gave comments to the papers and sent them to Editors. The Editors communicated with authors and reviewed again revised papers. The Editors made a final decision about acceptance of a paper.

- **Contact person for queries:** Anna Bogdan, email: coalise@tpu.ru

Content from this work may be used under the terms of the Creative Commons Attribution 3.0 licence. Any further distribution of this work must maintain attribution to the author(s) and the title of the work, journal citation and DOI.
Published under licence by IOP Publishing Ltd

Self-contained drilling assembly

L Yu Belozertseva

Chinakal Institute of Mining, Siberian Branch, Russian Academy of Sciences, Novosibirsk, Russia

E-mail: belozerceva@misd.ru

Abstract. Based on the analysis of technologies and equipment for the construction of borehole, the necessity and advantages of self-contained drilling assemblies are revealed and substantiated. The experience of engineering and application of such facilities is generalized. The key requirements to be imposed on a self-contained percussion drilling assembly are formulated. This paper can awaken interest of researchers and practitioners in mining machine engineering.

1. Introduction

The basic process in mineral exploration and mining is drilling. Drilling may be of various diameter, length and direction in rocks having different physical and mechanical properties. Advanced technologies in the mining industry impose new standards on drilling techniques and equipment, such as increased penetration rate, improved accuracy, borehole path changeability, lower weight and smaller size of drilling tools, operating cost saving, etc. [1, 2] To this effect, researchers persistently make every effect to improve and invent technologies and equipment for drilling.

2. A new and promising method of various diameter hole-making

A new and promising method of various diameter hole-making is drilling using machines equipped with a centerline for removal of drilling mud [3-5]. This method is advantageous for applicability in difficult geological conditions, under large inflows of water, in permafrost, or in exploration with core sampling. Another novel method offering high potential for drilling holes of large diameters to 3–6 m is using tangential–rotary picks was proposed at the Institute of Mining, SB RAS [6]. Evolvement and commercial-scale introduction of these methods can spur technological innovation in the mineral mining industry [7-10].

The routine method of drilling in present-day mineral mining is rotary–percussion drilling using drilling assemblies equipped with remote or down-the-hole pneumatic, hydraulic or another type drive. Rock breaking in this case takes place under the dynamic or static impact applied to the rock-breaking tool which is rotatable by the rock drill or drill pipe column. R&D activities aimed to improve rotary–percussion drilling equipment in Russia and abroad focus on finding and studying patterns of efficient rock–tool interaction [11-14], analysis of duty cycles of pneumatic [15–19] and hydraulic [20, 21] percussion machines, automation of drilling process, development of scientific framework for drilling automation engineering, as well as optimization of rock mass impact parameters depending on variable bottomhole conditions [22–26].

Advantages of the rotary–percussion drilling technologies are undeniable though large-size and expensive equipment sometimes prevents its application, for instance, in drilling in mine openings of

Content from this work may be used under the terms of the Creative Commons Attribution 3.0 licence. Any further distribution of this work must maintain attribution to the author(s) and the title of the work, journal citation and DOI.
Published under licence by IOP Publishing Ltd

small cross-section. In such cases, it is expedient to use self-contained drilling assemblies. Benefits of the latter include: avoidance of drill rigs with rotator and feeder, and, thereby, elimination of auxiliary operations associated with extension and tear-down of drill columns; remote control of the rodless drilling assembly by an operator located far from the hole mouth, which improves mining safety; lower price of the drilling equipment and operating cost saving due to the absence of a drill rig; simple handling of the drilling assembly in confined areas of mines (which allows using this equipment promptly in case of rescue in emergency situations); drilling of various-purpose holes in small cross-section underground openings, when the known drilling machines are low-efficient or inapplicable; controllability of a number of rodless drilling assemblies by a single operator, which raises productiveness of labor in drilling.

3. The rodless drilling assemblies

Application of rodless drilling assemblies is exemplified below.

a

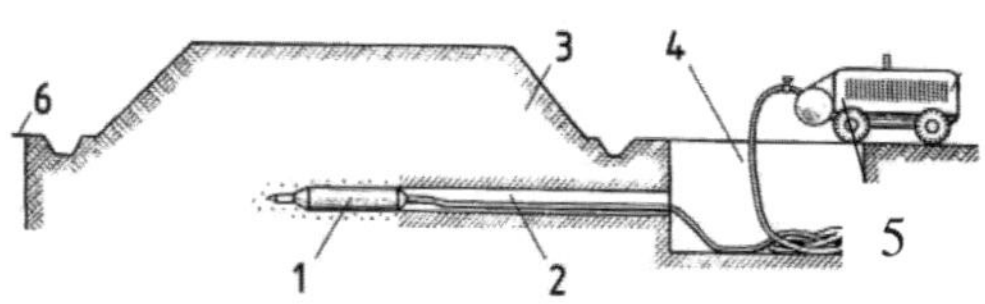

Vibro-impact punching: 1—air-driven puncher; 2—hole; 3—motor road; 4—process pit; 5—compressor; 6—outlet pit

c

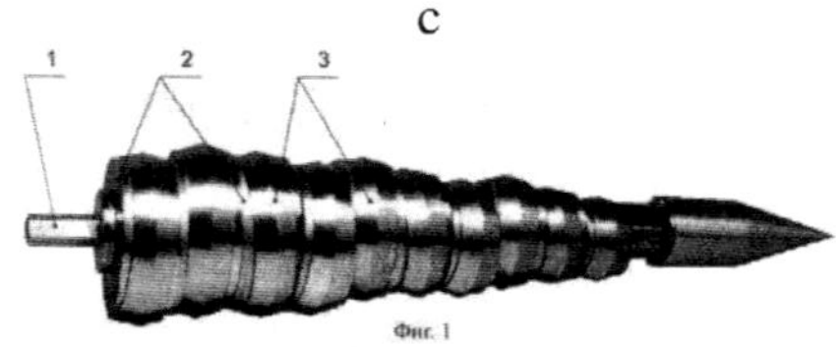

Hole-expanding tool

b

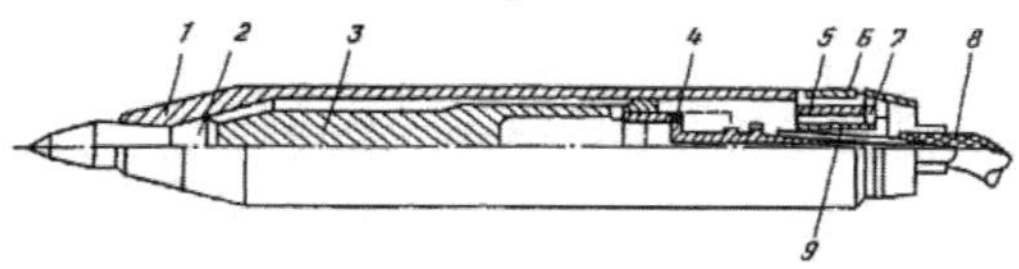

Air-driven puncher: 1—body; 2—anvil; 3—piston; 4—junction pipe; 5—damping device; 6—rear nut; 7—valve; 8—air feed hose; 9—nut

d

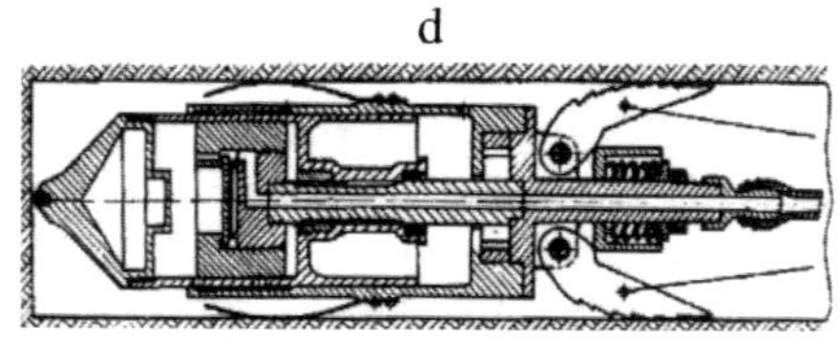

Rodless hammer drill

e

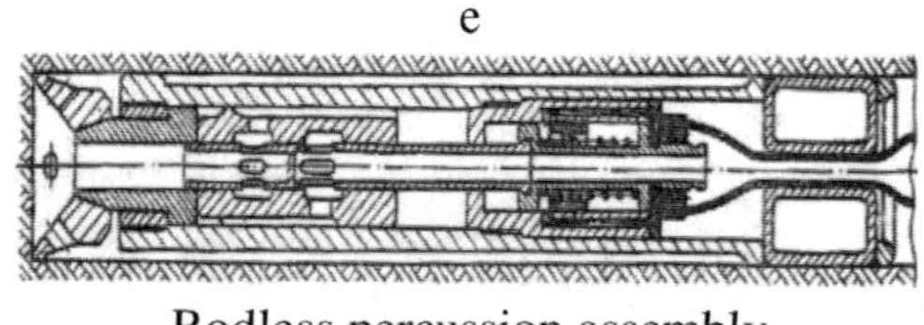

Rodless percussion assembly

f

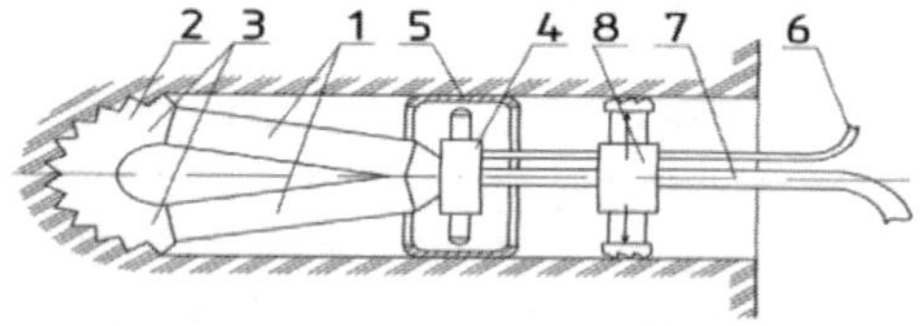

Adjustable path percussion drilling: 1—impactors; 2—rock-breaking tool; 3—sides of rock-breaking tool—4—switch; 5—slide bars; 6—cable (hose); 7—hose; 8—spacer

g

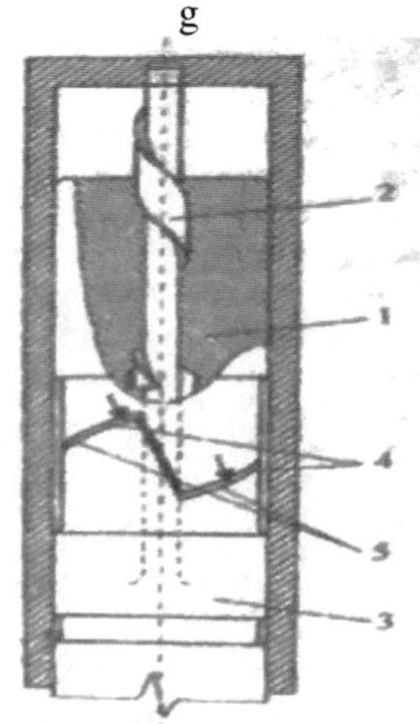

Longitudinally axial impact assembly

Figure 1. Rodless drilling technologies and equipment.

A widely spread method of drilling in compactable soil is vibro-impact punching (Figs. 1a and 1b). To this effect, it is possible to use air drills designed at the Institute of Mining, SB RAS. High capacity and reliability of these machines, as well as easy servicing and adaptability to confined site conditions are the major advantages of this equipment [27, 28].

The task tool of the air drills is a cylindrical body with inserted piston forced by compressed air to make to-and-fro motion and to hit the body. Under the impacts, the air drill body penetrates soil and makes a hole in it. Backward running of the machine is prevented by the friction forces on the side faces of the body. Thrust reverser enables backward motion.

Hole-making in compactable soil can also use soil drifters (Figure 1d). Such drilling facilities include a gearmotor, output and gear shafts, eccentric conical and cylindrical rollers, serially turned around the longitudinal axis of the drifter and capable of independent rotation, and sealer elements placed between the rollers. Shaft rotation makes the drifter move on in soil. The conical rollers make a hole and the cylindrical rollers expand and trim it (Figure 1c). The conformity of the helix leads, formed by points around which the rollers are turned, and the helix lead of roll of the rollers on the hole wall, as well as the cone angle of the rollers and the fitting conical helix leads stabilizes the tool and eliminates its slip in soil penetration.

For drilling in incompatible and strong rocks, an autofeed device is proposed for DTH drills [29]. This device has a front-end brake to prevent movement of the drill toward the bottomhole. The device is driven toward the bottomhole by the impacts applied to the rock-breaking tool and the body with brake. The brake represents a spring cutting cylinder which is pushed apart and remains in the hole when the device is removed from it. The same operating mode is a feature of the structural layout of the rodless DFH rotary–percussion drill (Figure 1d). Moreover, there is another back-end brake made as a row of sawtooth plates pressed to the hole walls by the center-located helical spring.

Inventor E.P. Varnello, Institute of Mining, SB RAS designed rodless air-hammering assembly BPP-1A [29] (Figure 1e) for drilling in coal and rocks having strength up to 30 MPa. The hammering piston inside the body moves to and fro under the action of compressed air and hits the rock-breaking tool and the front end of the body, which advances the assembly along the hole. The back-end brake prevents unwanted displacement of the body in the line of the bottomhole. Assembly BPP-1A was tested in mines. The assembly operated without any fault during the tests. Drill holes were cylindrically shaped and straight lined. Specifications of the drilling assembly are presented in Table 1.

Table 1. Specifications of rodless drilling assembly BPP-1A.

#	Description	Value
1	Drill hole diameter, mm	125
2	Drill hole length, m	50
3	Rated compressed air pressure, MPa	0.6
4	Blow energy, J	180
5	Blow frequency, Hz	11.1
6	Compressed air flow rate, m^3/s	0.135
7	Average penetration rate, m/s	$3.3\ 10^{-3}$
8	Assembly length, mm	1618
9	Assembly weight, kg	53
10	Starter weight, kg	32
11	Reverser	Available
12	Clean-out	Air
13	Hole inclination from the horizon	0–90

The Institute of Mining, SB RAS proposed also the method of adjustable path drilling (Figure 1f) [30, 31]. Here, there is a special rock-breaking tool, air/fluid or other source-driven hammering devices, setting tool and actuation/switch-over control. During drilling, impactor 1 alternately hits one of the sides of rock-breaking tool 2, which makes the tool turn in the force-and-aft plane and, thereby, break rocks on the bottomhole. At a certain turn angle, switch-over 4 actuates impactor 1 to hit the other side of rock-breaking tool 2. Fixtures 5 and 8 hold, fixture and press the tool. The operation control is implemented using hoses 6 and 7. The change in the drilling path is is possible through the special operation algorithm of the impactor.

In this drilling technique as against the other existing methods, penetrators of the rock-breaking tools deliver oblique impacts on rock, which considerably lowers energy input in the fracture process [32]. Furthermore, there is no need to rotate the rock-breaking tool around the hole axis, which saves from using a pipe column. The tool is pressed to the bottomhole and is advanced in the hole using a spacer. A drill rig is also withdrawn from the process, which considerably cuts down the cost of the equipment.

In terms of the self-sustained drilling assembly design, researchers E.P. Rusin and G.N. Khan from the Institute of Mining, SB RAS studied lengthwise rotary (screw) impact [33]. In this case, a hammering body moves progressively with simultaneous rotation around its own longitudinal axis. The screw impact is transmitted to a rock-breaking tool and, then, to rock. Rock is fractured in oblique percussion drilling, which is less energy-hungry and can enhance drilling efficiency at the same energy consumption [32]. It is possible to eliminate motor to rotate the assembly and, accordingly, the drilling column. As a result of the experimental research, a promising engineering solution and functional diagram were proposed for efficient screw impact (Figure 1g).

Hammer 1 and shank 2 of drill bit 3 make a nonself-stopping crew pair which carries out alternate motions with simultaneous leftward rotation. The colliding ends of hammer 1 and bit 3 are manufactured as screw surfaces 4 and 5, respectively, with opposite (rightward) rotation.

4. The engineering solution for the implementation of self-contained rodless percussion drilling assembly

Technological innovations in mineral exploration and mining impose new requirements on drilling technology and equipment. This calls for more extensive research in the field of self-propelling drilling assembly design and engineering. The currentness of the research is evident and indubitable.

Based on the review and analysis of the developed and tested drilling assemblies for soil and rocks, the design philosophy principles formulated for a new self-sustained percussive drilling assembly include: implementation of percussion drilling as the least energy-consuming method owing to efficient design of rock-breaking tool, with high potential to enhance drilling capacity; use of air, fluid or combination sources of energy for hammering and clean-out; advance and removal of the drill assembly by impacts, which eliminates the assembly jam while step-type feeders can cause jamming of the assembly in the hole, or its deviation from a preset direction; no in-built rotator, which essentially simplifies design and enhances operating reliability of the drilling assembly (it is necessary to find new engineering solutions to avoid an in-built rotator to make the rock-breaking tool rotate during drilling.

Based on the review and analysis of the domestic and foreign research, an engineering solution was proposed for the self-contained rodless percussion drilling assembly in conformity with the design philosophy principles above. The functional diagram of the assembly is shown in Figure 2.

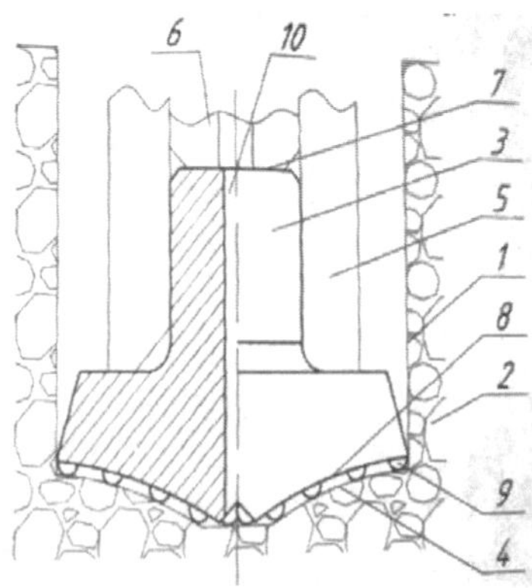

Figure 2. Functional diagram of percussion drilling assembly.

The assembly is composed of a percussion mechanism, an original-design rock-breaking tool with cutting inserts and a spacer (not shown in Figure 2) to prevent movement of the assembly in the direction from the bottomhole.

During drilling in hole 1 in rock mass 2, rock-breaking tool 3 is pressed to bottomhole 4. Impactor 5 is actuated and strikes with hammer 6 at end 7 of rock-breaking tool 3. The rock-breaking tool moves forward under the action of the blows and simultaneously rotates around its own axis in the transverse plane relative to hole 1. The rotation of the tool takes place due to the torque generated by the reaction forces of rock mass 2 in interaction with special spiral-shaped blades 8 equipped with cutting inserts 9. The spent energy source is used to clean bottomhole 4 and to remove drilling fines from hole 1 via channel 10.

The feature of this design is that rotation and progressive advance of the rock-breaking tool in the hole are implemented by the same hammering unit, which enables lower energy input in the drilling process. Damage takes place over the whole area of the bottomhole under oblique impacts of the tool.

Efficiency of the proposed drilling assembly, rock-breaking tool design and the tool–rock interaction parameters in drilling can be defined by physical modeling of rock destruction by impacts, which requires an original test-bench to be developed for the experimental studies. Such studies aimed to determine conditions for this percussion drilling technique, as well as to adjust geometry and energy of the new equipment are planned for the nearest future.

5. Conclusions

The author has reviewed the technologies and equipment in rodless drilling in soil and rocks. It is of the current concern to engineer self-propelling and self-contained drilling assemblies for medium and high strength rocks to ensure and advance technological innovations in mineral mining. The basic requirements imposed on the self-contained percussion drilling assembly being designed are formulated.

The engineering solution for the implementation of self-contained rodless percussion drilling assembly is proposed. The feature of the proposed design is that rotation and progression of the rock-breaking tool in a drill hole are executed by the same hammer source.

The avenues of the further research are identified to determine conditions of the proposed drilling method and to adjust geometry and energy characteristics of the novel drilling machine.

References

[1] Tao C, Cheng Y-F, Fang H, Yao Y-L, Wang Y-X, Ma H-H, Shen Z-W, Chen Y 2020 Fabrication and characterization of a novel underground mining emulsion explosive containing thickening microcapsules *Propellants, Explosives, Pyrotechnics* vol 45 **6** pp 932-941

[2] Šporin J, Vižintin G, Vukelić Ž, Mrvar P, Petrič M 2019 The characterization of wear in roller cone drill bit by rock material – sandstone *Journal of Petroleum Science and Engineering* vol 173 pp 1355-1367

[3] Zhang X, Luo Y, Gan X, Yin K 2019 Design and numerical analysis of a large-diameter air reverse circulation drill bit for reverse circulation down-the-hole air hammer drilling *Energy Science & Engineering* vol 7 **3** pp 921-929

[4] Lipin A A, Kharlamov Yu P and Timonin V V 2013 Circulation system of a pneumatic drill with central drilling mud removal *Journal of Mining Science* 49 2 pp 248–253

[5] Timonin V V and Kharlamov Yu P 2014 Mathematical modeling of circulation system operation in pneumatic hammer with center-line removal of drilling mud in Solidworks Flow Simulation *Vestn. KuzGTU* 6 (106) pp 61–65

[6] Timonin V V, Eremenko V A, Levenson S Ya, Lantsevich M A et al RF 2019 Patent RU 2709863 C1 Large-Diameter Drilling Method

[7] Barnov N G, Eremenko V A, Kondratenko A S and Timonin V V 2015 Substantiation of parameters of geotechnology for mining primary corundum deposits in difficult working conditions in upland areas *Gornyi Zhurnal* 11 pp 42–47

[8] Eremenko V A, Barnov N G, Kondratenko A S and Timonin V V 2016 Method for mining steeply dipping thin lodes *Gornyi Zhurnal* 12 pp 45–50

[9] Eremenko V A, Karpov V N, Filatov A P, Kotlyarov A A and Shakhtorin I O 2014 Improvement of ore blasting toward pre-broken rocks in rockburst-hazardous conditions *Gornyi Zhurnal* 1 pp 50–55

[10] Eremenko V A, Karpov V N, Timonin V V, Barnov N G and Shakhtorin I O 2015 Basic trends in development of drilling equipment for ore mining with block caving method *Journal of Mining Science* 51 6 pp 1113–1125

[11] Timonin V V 2008 Rock fracture by dynamic indentation of a group of indenters in nonlinear geomechanics, *Geodynamics and Stress State of the Earth's Interior: Conf. Proc. with Intern. Particip.* Novosibirsk pp 470–474

[12] Timonin V V 2009 Energy input in drilling by percussion machines *Mechanics of Modern Machines: Proc. 4th Int. Conf.* vol 2 pp 272–275
[13] Belousov A V and Timonin V V RF 2016 Patent RU 2535314 C1 Drill Bit
[14] Chanakya Nariseti 2013 Quantification of damage in selected rocks due to impact with tungsten carbide bits *Thesis Submitted Inconformity with the Requirements for the Degree of Master of Applied Science* Graduate Department of Civil Engineering University of Toronto
[15] Timonin V V 2015 DTH drill hammers for underground mining *Gorn. Oborud. Elektromekh.* 2 (111) pp 13–17
[16] Petreev A M and Primychkin A Yu 2015 Influence of air distribution system on energy efficiency of pneumatic percussion unit of circular impact machine *Journal of Mining Science* vol 51 **3** pp 562–567
[17] Petreev A M. and Primychkin A Yu 2016 Ring-type elastic valve operation in air-hammer drive *Journal of Mining Science* vol 52 **1** pp 135–145
[18] Lipin A A, Timonin V V, Tanaino A S 2006 *Modern DTH Percussion Machines for Drilling: Mining Equipment Catalogue–Handbook.* Saint-Petersburg pp 116–123
[19] Timonin V V 2009 Design of rock-breaking tool and hydraulic percussion machine for drilling in medium and high strength rocks *Synopsis of Candidate of Engineering Sciences Thesis* Novosibirsk: IGD SO RAN
[20] Gorodilov L V 2012 Analysis of the dynamics of two-way hydropercussion systems. Part I: Basic properties *Journal of Mining Science* vol 48 **3** pp 487–496
[21] Ma W, Geng X, Gao L, Liu Y, Tian X, Jia C 2019 Percussion characteristic analysis for hydraulic rock drill with no constant-pressurized chamber through numerical simulation and experiment *Advances in Mechanical Engineering* vol 11 **4** pp 835-842
[22] Kondratenko A S, Timonin V V, Karpov V N, Popelyukh A I 2018 Ways to improve rotary–percussive drilling efficiency *Gornyi Zhurnal* 5 pp 63–68
[23] Timonin V V, Karpov V N 2016 Assessment of rock failure process under percussion–rotary drilling, *J. Fundament. Appl. Min. Sci.* vol 3 **2** pp 172–176
[24] Oparin V N, Timonin V V, Karpov V N, Smolyanitsky B N 2017 Energy-based volumetric rock destruction criterion in the rotary–percussion drilling technology improvement *Journal of Mining Science* vol 53 **6** pp 1043–1064
[25] Karpov V N 2016 Assessment testing procedure for downhole pneumatic hammers under production conditions *J. Fundament. Appl. Min. Sci.* vol 3 **2** pp 74–80
[26] Karpov V N and Timonin V V 2016 Efficiency evaluation of DTH percussion machines in rotary–percussion drilling in mines *Integrated Subsoil Development and Conservation: Problems and Prospects. Proc. II Acad. Trubetskoy Int. School* Moscow: IPKON RAN pp 191–195
[27] Smolyanitsky B N, Oparin V N et al 2016 *Modern Technologies of Long-Hole Drilling in Rocks and Drilling Path Control Equipment* Novosibirsk: SO RAN p 342
[28] Kondratenko A S, Timonin V V, Abrov A A, Gosmanov M K, Esenov B U and Zharkenov E B 2014 Technology of safe trenchless horizontal and inclined drilling, *Vestn. KuzGTU* 1 (101) pp 40–45
[29] Varnello E P and Kamensky V V 1980 Functional diagrams and designs of rodless assemblies for drilling in rocks. Pneumatic percussion machines *Transactions of the Institute of Mining, SB AS USSR* Novosibirsk: IGD SO AN SSSR pp 3–15
[30] Svirshchevskii V K, Klimov N Yu RF 1995 Patent RU 2026936 C1 Hole-Expanding Tool
[31] Timonin V V, Alekseev S E and Chernienkov E M 2018 Patent RU2675614C1 Method of Shock Drilling
[32] Pokrovsky G N and Zakablukovsky N G 1976 *Rock Destruction by Oblique Blows with Regard to Drilling. Impact Transmission and Percussion Machines* Novosibirsk: IGD SO AN SSSR
[33] Rusin E P, Khan G N 2005 Surface form for transmission of longitudinal rotary impact *High Technologies of Mineral Mining and Processing: Proc. IV Conf. With International Participation* Novosibirsk: IGD SO RAN pp 226–231

Determination of rock fracture parameters in directional percussion drilling with drillhole path adjustability

E M Chernienkov

Chinakal Institute of Mining, Siberian Branch, Russian Academy of Sciences, 54, Krasnyi prospect, Novosibirsk, 630091, Russia

E-mail: e.chernienkov@misd.ru

Abstract: The paper reviews the existing technologies and equipment for drilling in rocks and soil with adjustable path of drillholes. The author describes a new method of percussion drilling and its application prospects. A lab-scale testing installation for physical simulation of rock fracture in drilling by the new method is presented. The force and impact characteristics required for the method implementation are updated.

1. Introduction

Drilling is an integral process of mineral mining. Development and improvement of drilling technologies and equipment will contribute to safer, easier and more efficient production of minerals [1, 2]. One of the promising and lively evolving trends in the drilling technology is controllable path drilling in very strong and medium-strong rocks.

This trend is initiated in the mining science to enhance efficiency of increasingly deeper level mining for higher quality delineation of ore bodies, better heading and shaft sinking using efficient drill and blast pattern designs at lower oversize yield [3, 4].

In the coal mining industry, controllable path drilling is applicable in top coal caving as adjustment of drilling paths can allow considerable reduction in work content of initiation slot drilling. Other than that, directional drilling technique simplifies drilling of service and drainage holes.

Potentially, adjustable path drilling is applicable in oil and gas reservoir engineering, when vertical drilling is complicated due to terrain conditions or can cause major environmental damage. Furthermore, the method can be effective in drilling with by-pass of undrillable zones or different facilities arranged on ground surface [5].

Thus, we need a technology for any-purpose production or shot long hole drilling in mineral mining at the ensured accuracy of path and with capacity to come to a preset point in underground space. Considering the above listed limitations, development of the controllable path drilling technology and equipment for very strong and medium-strength rocks is a relevant and challenging problem.

2. Controllable path drilling technologies and equipment

Hole-making with adjustable path in construction in very strong rocks uses the horizontal direction drilling technology (HDD) [6]. A HDD plant is equipped with special blades (figure 1). Drilling involves rotation of a drilling column and feed of an energy source, which is also a mud fluid. Intended deviation of the drilling path trajectory at a wanted angle is achieved by means of stoppage of rotation, which results in asymmetrical breakage of the hole bottom. As the required offset is gained, rotation is resumed

Content from this work may be used under the terms of the Creative Commons Attribution 3.0 licence. Any further distribution of this work must maintain attribution to the author(s) and the title of the work, journal citation and DOI.
Published under licence by IOP Publishing Ltd

with uniform cutting of the bottom hole and with straight-line drilling. The features of this method are difficult navigation of the drilling tool in underground space, as well as complex working environment, i.e. soil and rocks up to medium strength [7].

(a) (b)

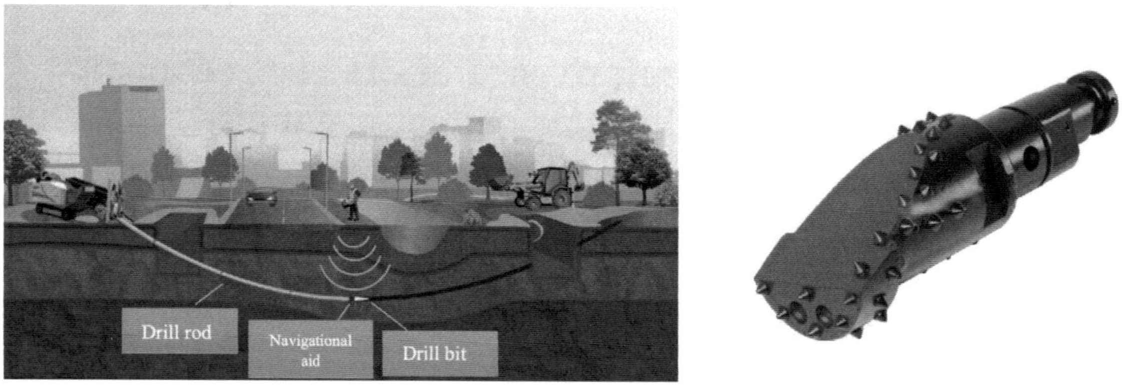

Figure 1. Horizontal directional drilling: (a) process flow chart; (b) customized drill blade.

In the known method of carving [8], a HDD plant is equipped with a DTH air hammer and a drill bit having an asymmetrical canted front. The bottomhole is cut uniformly at constant rotation of the drilling tool, and rerouting of the path is achieved via nonunform cutting of the bottomhole in alternate rotation of the drilling column at a certain angle. As against HDD with special blades, carving is applicable in a wider range rocks in terms of their strength though it also has a constraint which is associated with the need to prevent detorsion of the drilling column in reverse rotation, which greatly complicates the structural design and adds to the price of the equipment.

The oil and gas industries perform adjustable path drilling using rotary steerable systems and screw downhole motors with angle adjusters (figure 2).

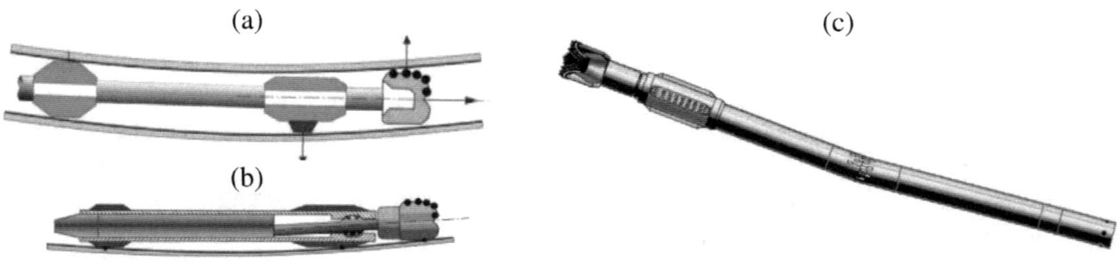

Figure 2. Rotary steerable systems and screw downhole motors with angle adjusters: (a) push-the-bit system; (b) point-the-bit system; (c) screw downhole motor with angle adjuster.

The rotary steerable systems bend drilling paths in three ways: Push-the-Bit—the assembly is pushed off the hole wall [9]; Point-the-Bit—the hole is bent through re-direction of the bit by bending the main shaft relative to the tool; Push-and-Point is a hybrid system combining the two methods above [10, 11].

Directional drilling with RSS most often uses structural configurations with the angle adjuster set between the drive and spindle assemblies with drive shaft running through. The wanted angle is set from the ground surface, by certain orientation of the lower adapter relative to the upper adapter. The range of the angle is from 0 to 2.5 deg.

Controllable path drilling in soil is performed using controllable air drills (figure 3). The drilling path is most commonly changed using differently shaped bits or by diversion of steering force [12-14].

Steering force is generated on air drills by pressure feed in the expanding cell. The cell expands and deflects the positioning elements which push the hole walls. The created steering effort changes the path of drilling. Shaft 2 can rotate in this case, and the process of rotary percussion drilling never stops.

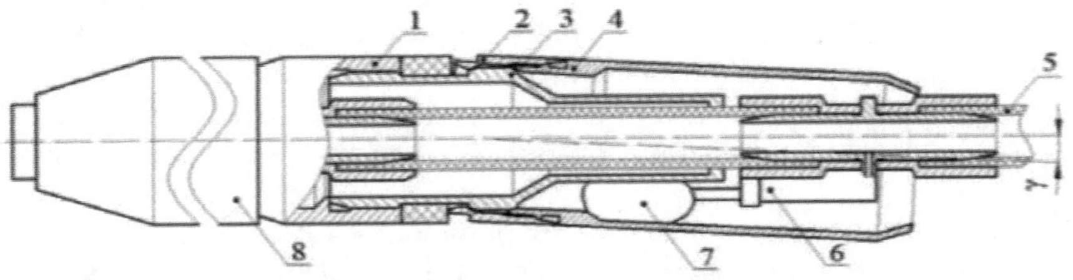

Figure 3. Flow chart of re-direction of air drill.

From the review of the known methods and means for controllable path drilling, we arrive at the conclusion that HDD, carving or RSS need a drill rig to transfer both commands and forces to the tool at the bottomhole via the drilling column. The use of large-size drilling installations is not always possible in the confined space of underground mines. For another thing, with drilling column in a drill hole, a lot of energy is lost in friction between the column and the hole walls and the column can break down. Accordingly, a promising trend in improvement of drilling equipment is engineering of autonomous rod-less drilling assemblies for controllable path drilling in medium and high strength rocks.

3. Controllable path percussion drilling

The Institute of Mining, Siberian Branch of the Russian Academy of Sciences has proposed a new approach to controllable path drilling [15]. Figure 4 illustrates schematically this method. Here, the main implementers are a special rock-breaking tool, air/fluid/or other source-driven percussion facilities, a spacer and a control system to actuate and switch over the percussion facilities.

During drilling, impactors 1 make alternate blows on the sides of rock-breaking tool 2. The percussion facilities may be DTH air drills and mechanism described in [16–18]. Under the action of alternate impacts, the rock-breaking tool rotates in the fore-and-aft plane and breaks the bottomhole. At a certain angle of rotation, switch 4 actuates an opposite impactor to strike the opposite the side of the rock-breaking tool. Holding down, fixturing and pressing of the tool is carried out by aids 5 and 8. Hoses 6 and 7 enable operation and control of the assembly. A special algorithm of operation of impactors allows the change in the drilling path.

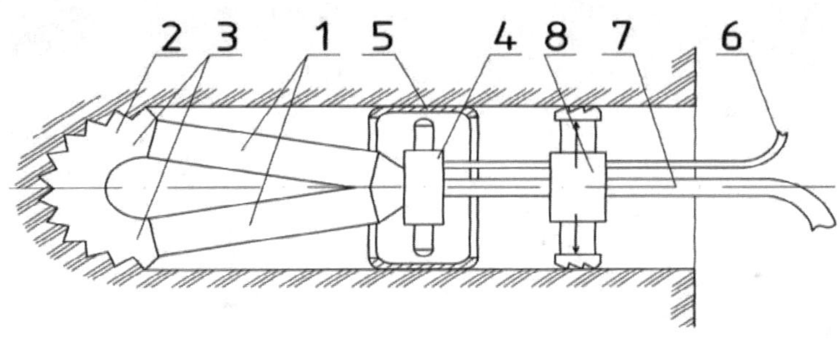

Figure 4. A new controllable path percussion drilling method: 1 - impactors; 2 - rock-breaking tool; 3 - sides of the rock-breaking tool; 4 - switch; 5 - slide bars; 6 and 7 - hoses; 8 - spacer.

This method differs from the known technologies by the oblique direction of blows which the impactors deliver to the rock-breaking tool, which reduces the energy intake of the process [19]. By varying the shape of the rock-breaking tool, it is possible to obtain holes of various cross-section: circular, square, rectangular, or slot-like. The other undeniable advantage of the new approach is unnecessity to rotate the rock-breaking tool around the drill hole axis, which eliminates drilling column from the equipment package. The tool is pressed to the bottom hole and is advanced by the spacer. Moreover, no large-size drill rig is required, which greatly saves the cost of the equipment.

Efficient operating mode of the drilling assembly to implement the least energy-consuming process of breaking in rocks of medium and high strength in case of straightline and directional drilling was determined after applied and basic research findings on rock fracture by the rock-breaking tool of the proposed shape and geometry under the joint effect of static and dynamic loads.

The tool–rock interaction was studied in physical simulation of the effect exerted on rock by the rock-breaking tool under static and dynamic loading on a specially designed testing installation.

4. Lab-scale testing installation for physical modeling of rock-breaking tool–rock interaction

The lab-scale testing installation in figure 5 allows physical simulation of rock fracture using the newly proposed method with simultaneous measurement of all parameters of interest (bend angle, rotation angle of rock-breaking tool, broken rock volume) and concurrent adjustment of parameters, which influence the process (hold-down of rock-breaking tool to bottom hole, unit blow energy, blow frequency, number of blows, etc.).

(a) (b) (c) (d)

Figure 5. Testing installation for physical simulation of rock-breaking tool–rock interaction: (a) general view; (b) hold-down gear; (c) shape of drill hole; (d) rock-breaking tool in the drill hole.

The testing installation in figure 5 is composed of a metal shape frame 1 and a rock mass model block 2. The test blocks can be of different size. Gear 3 to anchor and hold-down rock-breaking tool 4 to bottomhole is arranged on the frame. It was pre-set that the hold-down force was changed in the range from 50 to 300 N. At the top of the installation, there is a drop hammering facility 5 intended to deliver calibrated blows to two sides of rock-breaking tool 4. The range of the blow energy was 5–15 J. Before the tests started, an indent of the tool shape was made in in the block to eliminate pre-drilling, which was beyond the scope of the study, and to simulate the process of drilling. Penetration of the tool in the indent was recorded by a high-speed camera at a rate of 480 frames per second. After a series of blows on one side of the tool, broken material was removed from the hole scaled. The average angle of turn of the rock-breaking tool was determined as a ratio of the total angle of turn in the blow series to the number of calibrated blows. The limiting angle of turn of the rock-breaking tool of this size in the hole is 15 deg.

The rock mass model material was sand-and-cement with cement grade CEM I42.5 having the physical and mechanical properties as follows: compression strength 40–45 MPa and density 1800 kg/m^3. The block size was 400×400×400 mm. Such material simulates rocks of medium strength, is uniform, has no inclusions, can provide better repeatability of the experiment and makes it possible to adjust and approve the experimentation procedure initially.

First, a series of blows was made on one side of the rock-breaking tool at a certain blow energy and hold-down pressure. Then, the bottomhole was cleaned, chips were removed and scaled. The process was video recorded. After processing and interpretation of the test data, rock drilling efficiency in the new method will be assessed using the procedure from [20, 21] and compared with the conventional rotary percussion drilling technology [22].

After the first test series, it has been found that at the hold-down pressure of the rock-breaking tool to the bottomhole of 300 N, the tool jumps off the bottomhole and slips without damaging and cutting of the test material. Figure 6 presents the graph of the average tool turn angle versus its hold-down pressure. At the hold-down of 250–300 N and blow energy more than 5 J, the tool penetrates the block and drilling takes place. Thus, feasibility of the approach is proved, but it is required to modify the testing installation to implement higher axial hold-down pressure of the rock-breaking tool to the bottom hole and higher energy of unit blow. An increase in the hold-down results in a decrease in the average turn of the bit at fulfillment of penetration as the pressure neutralizes jump-off of the tool from the bottom hole.

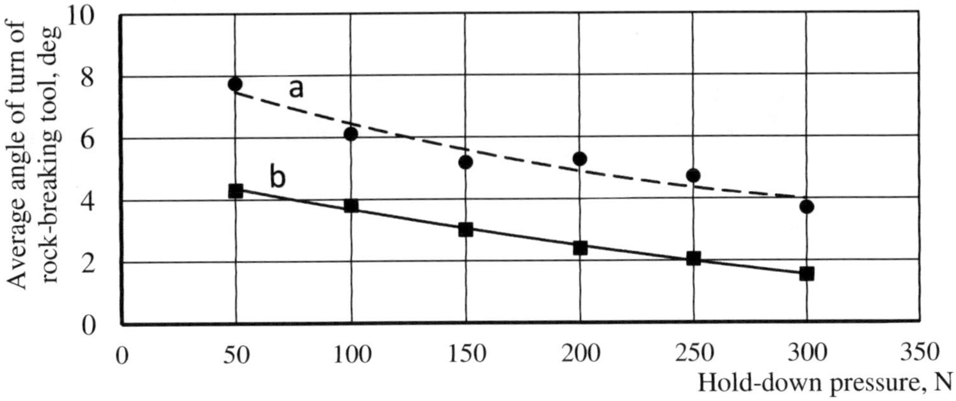

Figure 6. Average turn angle of rock-breaking tool versus its hold-down pressure to bottom hole: (a) blow energy 10 J; (b) blow energy 5 J.

5. Conclusions

1. The review and discussion of the existing controllable path drilling technologies and equipment for rocks and soil has revealed shortcomings of them.

2. The new method proposed for the directional drilling features adjustability of the drilling path and the capacity of the method to drill holes of differently shaped cross-section.

3. Feasibility of the new rock drilling method is proved on a laboratory scale. The force and impact values required for implementation of the new drilling method are refined, and appropriate modification of the testing installation is planned.

Acknowledgments
The reported study was funded by RFBR, project number: 19-35-90058/19.

References

[1] Karpov V N, Timonin V V, Konurin A I and Chernienkov E M 2019 Improvement of drilling efficiency in underground in mines in Russia, *IOP Conference Series Earth and Environmental Science* DOI: 10.1088/1755-1315/262/1/012024
[2] Kondratenko A S, Timonin V V, Karpov V N and Popelyukh A I 2018 Ways to improve rotary-percussive drilling efficiency *Gornyi Zhurnal* 5, pp 63–68
[3] Mametiev L E, Ananiev A N, Lyubimov O V and Zhaldin D V 2000 Prospects for horizontal drilling in underground mines, *Gorn Inform -Analit Byull*, 2000 11
[4] Levinson L M, Akbulatov T O and Akchurin Kh I 2000 *Borehole Bending Adjustment: Teaching Aid*, Ufa: UGNTU
[5] Norman J. Hyne 2001 Nontechnical Guide to Petroleum Geology, Exploration, Drilling and Production. 2nd Edition *Penwell Corp.*
[6] Serebrennikov A A 2020 *Laying Underground Engineering Communications with Horizontal Directional Drilling: A Training Manual* Moscow: KNORUS
[7] Kondratenko A S, Timonin V V and Patutin A V 2016 Prospects for directional drilling in hard rocks *Journal of Mining Science* 52 **1** pp 129–134
[8] Maksimov A S 2018 Improving the efficiency of a horizontal directional drilling rig in difficult soil conditions *Proc. II International Scientific and Practical Conference of Students, Graduate Students and Young Scientists* Omsk: SibADI pp 46–48
[9] Push-the-Bit RSS—Weatherford. Available at: https://www.weatherford.com/en/products-and-services/drilling/drilling-services/rotary-steerable-systems/push-the-bit-rss/
[10] Point the Bit Rotary Steerable Systems—Schlumberger. Available at: https://www.slb.ru/
[11] Point+Point— Halliburton. Available at: https://www.halliburton.com/
[12] Danilov B B, Smolyanitsky B N and Cheshchin D O 2019 Design parameters of the device for adjusting a trajectory of pneumatic puncher motion in soil *IOP Conf. Ser.: Earth Environ. Sci.* 262 012012
[13] Danilov B B, Smolyanitsky B N, Chanyshev A I and Cheshchin D O 2018. Determination of pneumatic puncher turn radius during change of its motion path in solid *Journal of Mining Science* 54 **3** pp. 397–403
[14] Danilov B B, Smolyanitsky B N and Cheshchin D O 2015 Justification of basic diagrams of horizontal drilling deflectors *Journal of Mining Science* 51 **3** pp 553–561
[15] Timonin V V, Alekseev S E, Chernienkov E M 2018 Patent US2675614A Method of shock drilling
[16] Timonin V V DTH air hammers for underground mines. *Gorn. Oborud. Elektromek.* 2(11) pp 13–17
[17] Timonin V V, Alekseev S E, Kokoulin D I, Kubanychbek and Chernienkov E M 2018 Air distribution analysis in DTH air hammer with valve–piston couple *Interexpo Geo-Sibir Conf. Proc.* 6 pp 207–217
[18] Timonin V V, Alekseev S E, Kokoulin D I, Kubanychbek B and Chernienkov E M 2018 Influence of part-to-part gaps in air distribution system on operation of DTH hammer *IOP Conference Series Earth and Environmental Science* DOI: 10.1088/1755-1315/262/1/012073

[19] Pokrovsky G N and Zakablukovsky N G 1976 Analysis of rock fracture process under oblique impact as applied to drilling *Transfer of Impact and Percussive Machines* pp 51–65 (in Russian)
[20] Timonin V V and Karpov V N 2016 Assessment of rock failure process under percussion–rotary drilling *Fundament. Prikl. Vopr. Gorn. Nauk* vol 3 **2** pp 172–176
[21] Karpov V N, Timonin V V, Tkachuk A K, Konurin A I et al 2018 Patent RU2674350C1 Acoustic test of impact of percussion tools of mining and construction machines on geomedium
[22] Oparin V N, Timonin V V, Karpov V N, Smolyanitsky B N 2017 Energy-based volumetric rock destruction criterion in the rotary–percussion drilling technology improvement *Journal of Mining Science* vol 53 **6** pp 1043–1064

IOP Conf. Series: Materials Science and Engineering 1118 (2021) 012003 doi:10.1088/1757-899X/1118/1/012003

Investigation of the Criteria for Evaluating Electrode Metal Transfer in Short Circuit Gas-Shielded Arc Welding

S V Bolotov

Faculty of Electrical Engineering, Belarusian-Russian University, 43, Mira Ave, Mogilev, 212000, Belarus

E-mail: s.v.bolotov@mail.ru

Abstract. Based on the developed information-measuring complex, the investigation of processes of melting and electrode metal transfer during short circuit arc welding was carried out. The close relationship between the energy of the arc at the stage of its formation and the diameter of an electrode metal droplet was found. It is proposed to evaluate the transfer stability by the harmonic composition of welding current.

1. Introduction

The strength properties of welded joints are closely related to the processes of heating, melting and transfer of electrode metal droplets. The type of transfer is determined by the parameters of welding conditions [1]. In consumable electrode gas-shielded arc welding, the short circuit transfer of droplets to the weld puddle has become widely used and it has a number of technological advantages such as deep penetration, high productivity and high-quality weld formation in any spatial position [2].

The study of electrode metal transfer in arc welding is relevant to the recording of fast processes, since the frequency of droplet separation can reach 300 Hz. Complexes for studying droplet transfer during welding are unique equipment and they make it possible to synchronously playback frames of the recorded welding process and oscillograms of the welding current and voltage [3,4]

2. Equipment and research methods

An information-measuring complex for investigating the processes of melting and electrode metal transfer in arc welding was developed at the Belarusian-Russian University [5]. The overall view of the complex is shown in Figure 1.

The movement of the welding torch at a programmable speed is carried out using the FANUC Robot ARC Mate 100iC industrial robot. The Fronius TransPuls Synergic 3200 is used as a welding power source, which allows performing consumable electrode gas-shielded arc welding using a 0.8 ... 1.6 mm diameter welding wire over a welding current range of 3 to 320 A (wire feed speed of 0.5 – 22 m/min) at the arc voltage of 14 – 30 V. The welding table is installed on an optical bench. To illuminate the welding zone and eliminate powerful light emission of the welding arc, a 3W 850nm LED was used supplied with 700 – 800 mA current from the APS-1721L source. The selection of the required radiation spectrum was carried out using the IKS6 infrared filter (850-3,000 nm) and the NS13 neutral light filter (to weaken the brilliance of radiation emitted by red-hot objects). The Evercam 1000-4-C with a CMOS sensor was used as a high-speed camera with frame rates between 1,000 and 4,000 fps and the minimum exposure time of 1 μs. The recording of welding process

Content from this work may be used under the terms of the Creative Commons Attribution 3.0 licence. Any further distribution of this work must maintain attribution to the author(s) and the title of the work, journal citation and DOI.
Published under licence by IOP Publishing Ltd

parameters, such as instantaneous values of welding current and arc voltage, and shielding gas consumption, was carried out by the RSP-BRU-2001 recorder developed at the Belarusian-Russian University. The NI USB-6009 with a sampling rate of 20 kHz, the LT 1000-SI current sensor (1...1,000 A), the LV 25-P voltage sensor (10...500 V), and the AWM5104 shielding gas flow meter (0...20 l/min) were used as a data acquisition unit of the recorder. The complex makes it possible to perform welding in the modes set by the operator, to record and measure the parameters of welding processes with a synchronous superposition of frames recorded with the high speed camera on the obtained oscillograms.

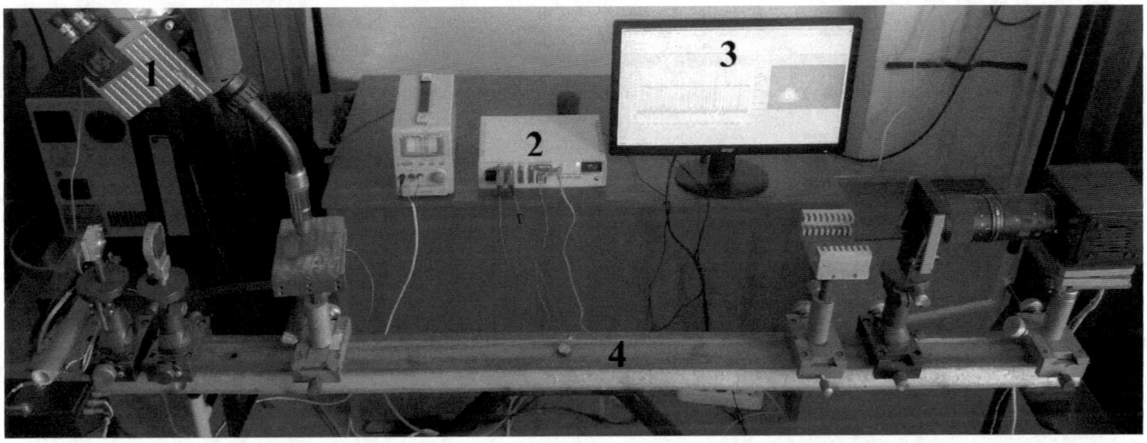

Figure 1. The information- measuring complex: 1 – device for automatic welding torch movement, 2 – welding process recorder, 3 – personal computer, 4 – optical bench with a high-speed video camera.

The software to control the information-measuring complex was developed in the NI LabVIEW graphical programming environment [5].

To study the criteria for assessing the electrode metal transfer in short circuit arc welding, a series of welding deposition experiments was performed on 100x100x10 mm plates made of low-carbon steel using a 1.2 mm diameter Sv08G2S wire in an 82% Ar + 18% CO_2 gas mixture. The welding speed was 18 m/h with the electrode protrusion of 14 mm and shielding gas consumption of 14 l/min. The experiments were carried out at the welding current values ranging from 130 A to 200 A, with a stepwise arc voltage change from 16 to 24 V. The main criteria for evaluating the welding process were programmatically determined for each microcycle of the electrode metal transfer (Figure 2). The determined criteria are as follows: the frequency of droplet short circuits f_{cycle}, the duration of arc burning t_{arc}, the duration of short circuit t_{shot}, the maximum short circuit current I_{peak}, the rate of its rise dI_{peak}/dt, as well as the arc energy Q spent to form a drop of molten metal. These criteria were compared with the diameter of the drop d_{drobe} at the moment of the short circuit with the weld puddle.

3. Research findings and their discussion

The frequency of droplet short circuits f_{cycle} is considered to be the main criterion for evaluating the welding process [6-8]. The typical values of the cycle frequency in short circuit welding are 5 – 200 Hz [2].

Figure 2 shows the oscillograms of welding current *i*, arc voltage *u*, welding power *p* and the frames of high-speed recording during one microcycle of one drop transfer corresponding to points 1, 2, 3 on the graphs. At the time point corresponding to point 1, the arc is ignited, while the current attains a maximum value of I_{peak} = 370 A. As the electrode wire is fed and melted, the drop diameter increases (point 2), while the welding current and arc voltage fluctuate within insignificant limits. At the moment when a molten electrode metal droplet develops the short circuit with the weld puddle (point 3), the voltage tends towards zero due to the overlap of the interelectrode gap, and the welding current begins to increase to a maximum value. A gradual transfer of the droplet into the weld puddle occurs during t_{shot} = 4 msec, and after the rupture of a liquid bridge, the re-excitation of the arc takes place. The complete cycle of one drop transfer is T_{cycle} = 68 msec (frequency f_{cycle} = 1/ T_{cycle} = 14.7 Hz) with the drop diameter d_{drob} = 2.7 mm, and arc burning time t_{arc} = 64 msec.

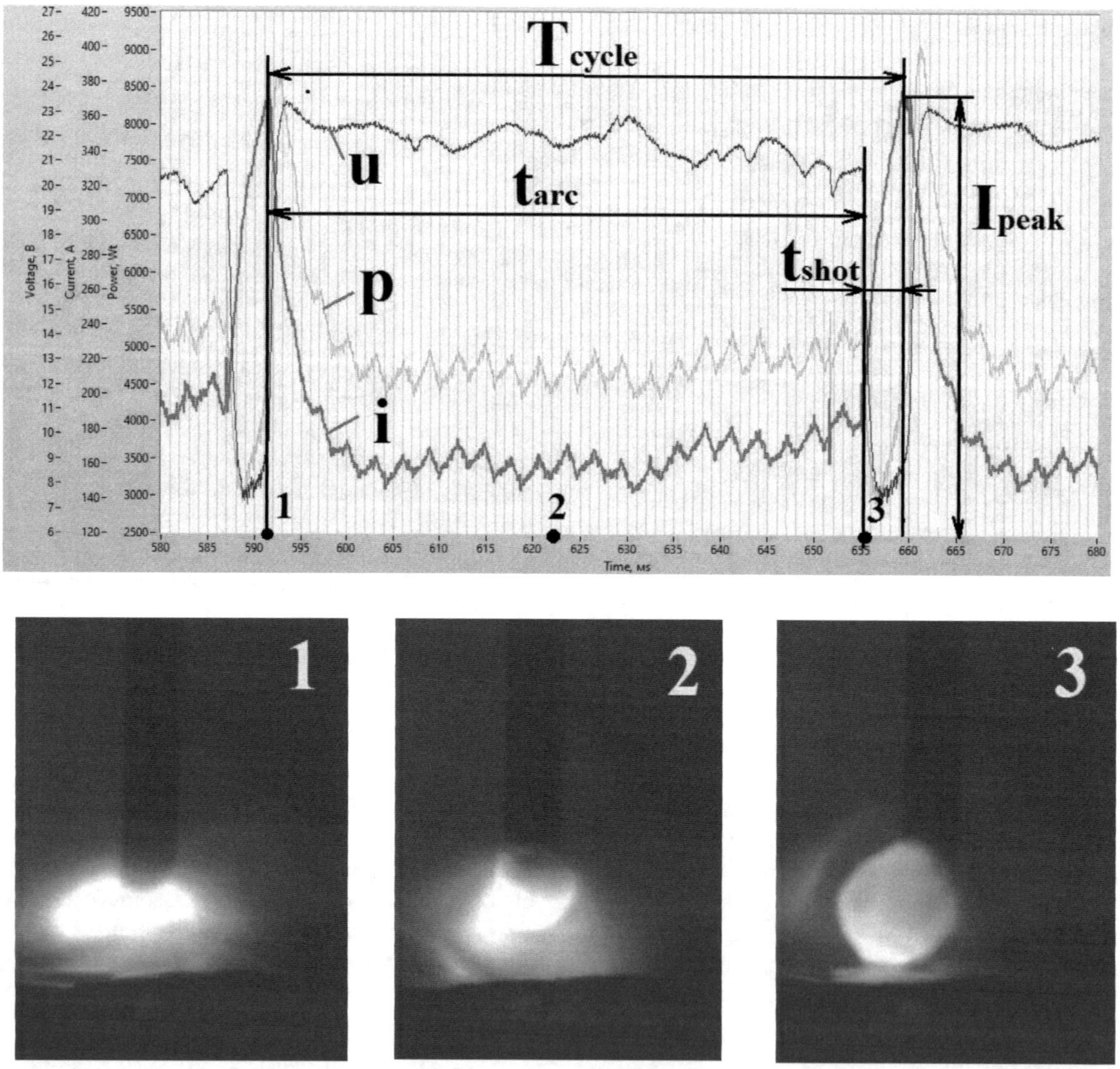

Figure 2. Oscillograms and frames of video shooting of short circuit electrode metal transfer.

Over the course of research, it was found that an increase in the welding current (wire feed speed) leads to an increase in the frequency of short circuits (table 1, figure 3). With an increase in the arc voltage, the frequency of the cycle of short circuits decreases, which is due to an increase in the interelectrode gap and an increase in the droplet size. At low arc voltage, the frequency of short circuits decreases faster.

Table 1. Results of the research into the criteria for evaluating the arc GMAW process.

I, A	U, V	Δf, Hz	f_{cycle}, Hz	I_A, A
130	16	45-59	52,5	8,6
130	17	32-46	39	9
130	18	23-32	28	20,3
130	19	21-22	21,2	21,6
130	20	8-18	12,5	13,2
130	21	5-20	10	8,3
200	18	35-51	43,5	7,5
200	19	24-34	29	8,3
200	20	16-23	19	9,5
200	21	13-17	15	10,1
200	22	11,5-13	12,7	10,5
200	23	9-12	10,5	6,5
200	24	5-14	9,5	5,7

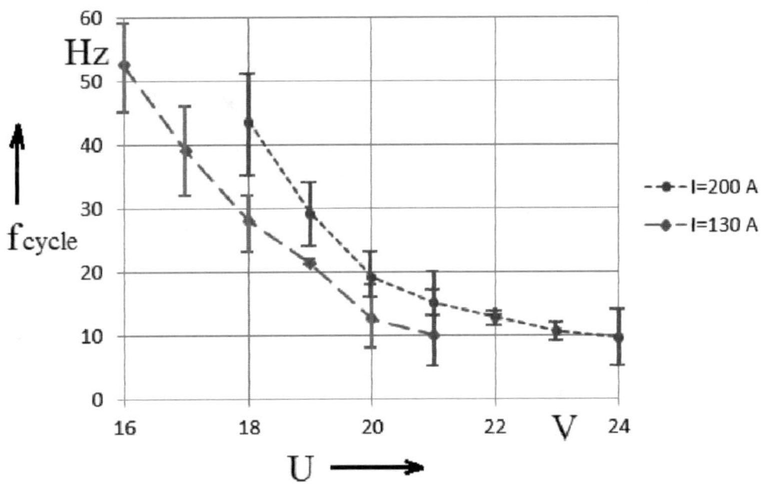

Figure 3. The relationship between the frequency of the cycle of short circuits and the parameters of the welding mode.

It was found that short circuiting of the arc gap does not always cause drop transfer, and when determining the transfer frequency, short circuits with the duration of t_{shot} <1.6 msec should not be taken into account. The relationships between the investigated criteria for evaluating transfer and the diameter of electrode metal droplets were obtained. It was found that the energy of the arc at the stage of its formation has the closest relationship with the droplet diameter (Figure 4). The regression equation will be

$$d_k=1,63+0,0566\cdot Q$$

The correlation coefficient is 0.98. The resulting relationship indicates that when the arc energy changes from 2.7 to 22.6 J, the diameter of an electrode metal droplet formed, changes from 1.8 to 2.9 mm. This linear relationship is valid when a Sv08G2S welding wire of 1.2 mm diameter is used in a 82% Ar + 18% CO_2 mixture in the welding modes that provide short-circuit transfer of the arc gap, and it can be used in the systems for automatic control of welding processes to dose energy needed to form drops of the desired size.

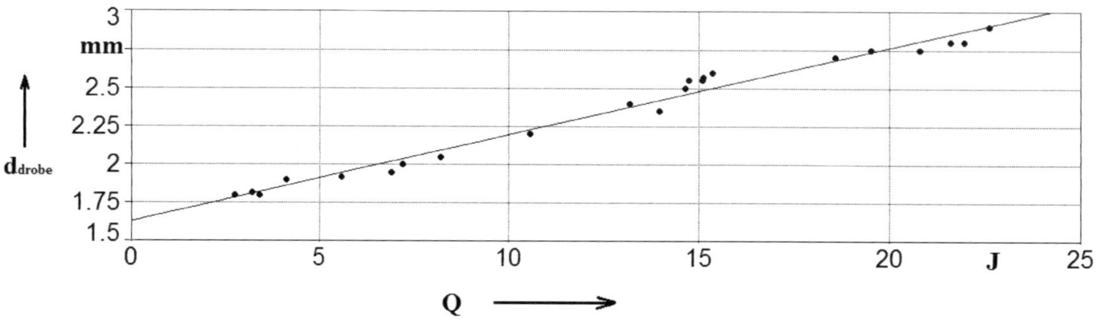

Figure 4. Relationship between the diameter of an electrode metal droplet and the arc energy.

It is proposed to evaluate the stability of electrode metal transfer by the harmonic composition of the welding current in the 5 – 200 Hz range determined by the software spectrum analyzer (Figure 5). Thus, at 130 A welding current and 19 V arc voltage, a stable process of the transfer of electrode metal droplets with a frequency of 21 Hz is observed, which corresponds to the maximum amplitude of the welding current harmonic at the same frequency.

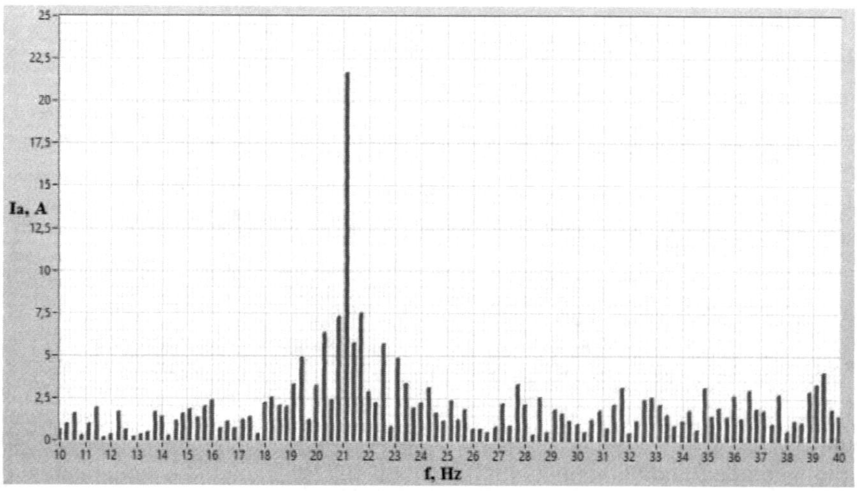

Figure 5. Harmonic composition of the welding current for a stable process.

It was found that for a stable welding process, the current frequency corresponding to the fundamental frequency of electrode metal droplet transfer has the maximum value of amplitude I_a (Table 1). The width of the spectrum with current amplitudes commensurate with the amplitude of the fundamental frequency does not exceed 1 – 3 Hz. With an increase in the value of the welding current from 130 to 200 A, the amplitude of fundamental harmonic of current I_a, corresponding to the fundamental frequency of droplet short circuits, decreases from 21.6 A to 10.5 A (Figure 6).

The obtained relationships can be used to assess the stability of the arc welding process with short circuits of the arc gap.

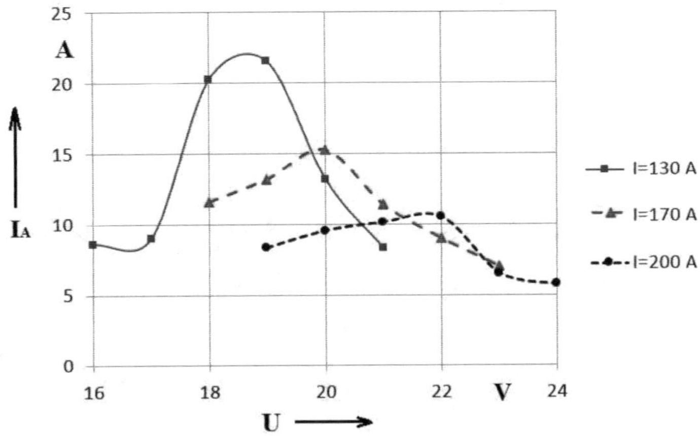

Figure 6. The relationship between the amplitude of the current of fundamental harmonic and the parameters of the welding mode.

4. Conclusions

The information-measuring complex was developed to study the processes of melting and electrode metal transfer in arc welding.

It was discovered that the energy of arc at the stage of its formation has the closest relationship with the diameter of an electrode metal droplet, when short circuit welding is performed using a 1.2 mm diameter Sv08G2S wire in an 82% Ar + 18% CO2 shielding gas mixture within the 130 – 200 A current and 16 – 24 V arc voltage range.

It is proposed to evaluate the stability of the arc welding process based on the harmonic composition of the welding current in the 5 – 200 Hz range determined by the software spectrum analyzer. It was found that for a stable welding process, the current frequency corresponding to the fundamental frequency of the electrode metal droplet transfer has the maximum value of the spectrum amplitude, which decreases as the current increases.

References

[1] Potapievsky A G, Saraev Y N and Chinakhov D A 2012 Consumable Electrode Shielded-Gas Welding. Equipment and Technology of the Future *Tomsk Polytechnic University Publishing House* p 208

[2] Milyutin V S 2019 Testing of Welding Properties of Arc Welding Equipment (Ekaterinburg) p 466

[3] Rhee S 1992 Observation of metal transfer during gas metal arc welding *J Weld. J.* **10** 381–386

[4] Saraev Y N, Lunev A G, Kiselyov A S, Gordynets A S and Trigub M V 2018 Complex for investigation of arc welding processes. *J. Aut. Weld.* **8** 15–24

[5] Bolotov S V, Khomchenko A V, Shulga A V and Bolotova E L 2020 Information-measuring complex for investigation of melting and electrode metal transfer in arc welding *J. Bryansk State Tech. Univ. Bulletin* (Bryansk: BGTU) **6 (91)** 4–11

[6] Lenivkin V A, Dyurguerov N G and Sagirov H N 2011 Technological Properties of Gas-Shielded Arc *Moscow: Gefest* p 368

[7] Pan J 2003 Arc Welding Control *England: Cambridge Woodhead Publishing Limited* p 604

[8] Gladkov E A, Yushin A A and Perkovsky R A 2011 Assessment of welding properties of equipment with inverter power sources based on energy characteristics *J. Welding and Diagnostics* **1** 31–35

IOP Conf. Series: Materials Science and Engineering 1118 (2021) 012004 doi:10.1088/1757-899X/1118/1/012004

Ion-Plasma Treated Parts Quality Improvement Analysis Based on the Reliability Theory Criteria

V I Bogdanovich and M G Giorbelidze

Samara National Research University, 34, Moskovskoe Highway, Samara, 443086, Russian Federation

E-mail: bogdanovich@ssau.ru; m.giorbelidze@ssau.ru

Abstract. The procedure of using reliability theory indices to evaluate parts vacuum ion-plasma treatment efficiency has been considered. Efficiency coefficient of technological procedures improving life and reliability of engineering parts taking into account alteration of average index value, its dispersion, and probability level of faultless operation has been suggested.

1. Introduction

Efficiency of procedures on reliability and quality improvement of parts by treatment is usually evaluated comparing average values of the given quality coefficient of a part with a coating and without it. However, this approach is not complete, as in some cases not only average index value, but the dispersion also changes. As a result, complete analysis based on criteria of reliability theory of part quality improvement with ion-plasma treatment is of interest. Moreover, methodology of reliability theory predicts significant impact of dispersion parameter on quality coefficients [1].

Actually, any coefficient of product quality formed when implementing a technological method is a random variable complied with the certain distribution law. This distribution is usually characterized by two numerical characteristics - mathematical expectation (average value) and dispersion (mean square deviation). Naturally, it is impossible to compare quantitatively two distributions of the quality coefficient values before and after treatment, but it is acceptable to be restricted by comparison of only average values at close values of their dispersions [1].

At the same time, in reliability theory, indices of ‘faultless probability’ and ‘g-percentile lifetime’ for objective evaluation are introduced. They compare indices with various dispersions, but they are not widely used in technical literature.

2. Formulation of the mathematical model for determination of performing the task probability

Let us consider methods of applying of these indices of the reliability theory to evaluate efficiency procedures on vacuum ion-plasma treatment of parts. This method is connected with the task of determination of probability of the task performing [1] that can be formulated with the following way. Let us assume a parameter of quality Y is a random variable distributed according to the normal law with parameters m_y (mathematical expectation) and σ_y (dispersion). When using, parts are acted upon by external factors. Their evaluation gives limiting permissible value for the considered quality

Content from this work may be used under the terms of the Creative Commons Attribution 3.0 licence. Any further distribution of this work must maintain attribution to the author(s) and the title of the work, journal citation and DOI.
Published under licence by IOP Publishing Ltd

parameter. In general case, this limiting value X is a random variable. This random variable due to the central limiting probability theorem would be distributed according to the normal law with parameters m_x and σ_x. Then faultless operation on quality parameter Y with probability γ would be determined by the inequality:

$$P(Z = Y - X \geq 0) \geq \gamma . \quad (1)$$

In special cases, value X can be a deterministic constant (limit value of an external stress, technological tolerance, permissible alteration of a parameter y, etc.); inequality signs can be inverse, values Y and X can be pre-failure lives or parameter values that vary with time, etc.

Since Y and X are subjected to Gauss distribution, their composition $Z = Y - X$ would be also subjected to Gauss distribution with parameters $m_z = m_y - m_x$ and $\sigma_z^{\ 2} = \sigma_y^{\ 2} + \sigma_x^{\ 2}$, and probability of meeting condition (1) for the extreme case $P(z) = \gamma$ and $z = 0$ would be determined by the integral

$$\gamma = P(z=0) = 0.5 - \Phi\left(U_\gamma(z=0)\right), \quad (2)$$

where $\Phi(U_\gamma) = \frac{1}{\sqrt{2\pi}} \int_0^{U_\gamma} \exp\left(-\xi^2/2\right) d\xi$ – error integral, $U_\gamma(z) = -\frac{m_z}{\sigma_z}$ – quantile of Gauss distribution for some probability γ.

Let us transform the expression for the quantile $\left(U_\gamma(z=0)\right)$ to the following

$$U_\gamma(z) = -\frac{n-1}{\sqrt{n^2 v_y^2 + v_x^2}}, \quad (3)$$

where $n = m_y / m_x$ – safety factor (reliability margin); $v_x = \sigma_x / m_x$, $v_y = \sigma_y / m_y$ – coefficients of variations.

When value X is a deterministic constant with the value $X = X_0$, expression for the quantile (3) is the following

$$U_\gamma = -\frac{n-1}{n v_y}, \quad (4)$$

where $n = m_y / X_0$.

Before total analysis of equations (1-4), let us consider the example from practical experience of vacuum ion-plasma deposition [2-10].

3. Tests of samples with a coating and without it

Upon implementation of vacuum ion-plasma deposition technology of coatings based on titanium nitride to protect a part made of titanium alloy, the goal was set to discover influence of these coatings on endurance limit. In this connection, comparison tests of standard samples made of titanium alloy with coatings and without them were carried out. Next, a lot of compressor rotor blades were tested.

Tests were carried out on smooth samples with round cross-section according to the scheme of pure bending under rotation up to complete destruction or 10^7 cycles. The following results have been obtained: for the samples without a coating $\bar{\sigma}'_{-1} = m_y' = 530$ MPa, $\sigma_y' = 58$ MPa; for the samples with a coating $\bar{\sigma}_{-1} = m_y = 557$ MPa, $\sigma_y = 45$ MPa.

Since the values m_y' and m_y differ insignificantly (5 %), the standard conclusion has stated that ion-plasma coatings obtained at the given mode do not decrease endurance limit. This fact has resulted in their further implementation (naturally, on substantial increase of wear resistance of the system). But for the samples with a coating and without it, dispersion magnitudes of average values differ in 1.29 times. That is why let us calculate probabilities of faultless operation for two cases.

On the assumption that safety factor for the sample without a coating is chosen $n' = 1.50$, for the samples with a coating we get $n = 1.58$. Coefficients of variations are $v_y' = 0.11$ and $v_y = 0.08$, so quantiles (3) would be $U_\gamma' = -3.03$ and $U_\gamma = -4.59$ for the samples without coatings and the samples with coatings respectively. Using the tables of error integral and according to (2) we would obtain the values of faultless operation for the sample without a coating $\gamma' = 0.9987$ and with a coating $\gamma = 0.9999$, so failure probability ($1-\gamma$) has been decreased in $(1-\gamma)/(1-\gamma) = 130$ times.

Thus, efficiency of a vacuum ion-plasma coating can be evaluated as decrease of failure probability in 130 times because of fatigue, though average value of endurance limit has increased only by 5 %. However, it is obvious that in most cases there is no need to provide computational probability of faultless operation at the level of 'five nines'. That is why test results of the samples with and without coatings are to be reduced to the single reasonable value of faultless operation probability.

One possible variant of implementation of efficiency coefficient of conducted technological procedures is an approach similar to implementation and comparison of g-percentile lifetimes. Let us consider the leading example. When testing hard-alloy plates with a multilayer ion-plasma coating and without a coating in the delivery condition, the following results have been obtained for average pre-failure life and their dispersions $m_t = 33.7$ min, $m_t' = 22.5$ min, $\sigma_t = 6.7$ min, $\sigma_t' = 7.9$ min. Then, to complete the technological task with probability $\gamma = 0.99$ it is necessary to prescribe g-percentile lifetimes (efficient life) of instrument use according to (4):

$$T_\gamma = m_t\left(1+U_\gamma v_t\right);\; T_\gamma' = m_t'\left(1+U_\gamma v_t'\right), \tag{5}$$

where in (4) it has been redesignated $X_0 = T_\gamma$; $X_0' = T_\gamma'$; $m_y = m_t$; $m_y' = m_t'$; $\sigma_y = \sigma_t$; $\sigma_y' = \sigma_t'$

For the chosen probability of completing the technological task $\gamma = 0.99$ we get $T_\gamma = 18.0$ min and $T_\gamma' = 4.2$ min. Thus, though increase of average pre-failure life of the instrument with the coating has increased in $m_t / m_t' = 1.5$ times, real increase of efficient instrument life has increased in $T_\gamma / T_\gamma' = 4.3$ times. Consequently, actual efficiency of the quality parameter (efficient life) is evaluated by the coefficient that is equal to:

$$\alpha_{ef} = \frac{T_\gamma}{T_\gamma'} = \frac{m_t\left(1+U_\gamma v_t\right)}{m_{t'}\left(1+U_\gamma v_{t'}\right)}. \tag{6}$$

4. Evaluation technique of a technological procedure on permissible g-percentile alteration of quality coefficient

Returning to the example with the endurance limit, we have suggested to evaluate a technological procedure on permissible g-percentile alteration of quality coefficient. In this case, similar to (5) and (6) efficiency coefficient of the technological procedure on the quality parameter *Y* is in the form of:

$$\alpha_{ef} = \frac{m_y\left(1+U_\gamma v_y\right)}{m_{y'}\left(1+U_\gamma v_{y'}\right)}. \tag{7}$$

Calculation according to equation (7) is carried out the following way. First, mathematical expectation m_y' and standard deviation σ_y' of quality coefficient *Y* are determined. After the technological procedure these quantities take values m_y and σ_y. Using these values and the given value of probability of faultless operation γ on this parameter, we calculate according to equation (7) (value of a quantile U_γ is determined using standard tables of error integral). For example, in the above considered example on endurance limit increasing $\alpha_{ef} = 1.20$ at $\gamma = 0.9987$ and $\alpha_{ef} = 1.37$ at $\gamma = 0.9999$, that is procedure efficiency gives index increase by 19 % and 37 % rather than 5 %. In this case, it is clear that quality coefficient (7) also objectively signifies quality increase of a part on endurance due to deposition of vacuum ion-plasma coatings. Besides, in case of requirements raise to a part on the value of probability of faultless operation, efficiency of ion-plasma technology increases.

At the same time, particular interest is the strategy of increase of faultless operation probability on the given quality parameter. From (1) and (2) it is evident that increase of faultless operation probability $P \to 1$ is connected with necessity to provide rise of $|U_\gamma|$. Rise of $|U_\gamma|$ can be provided by the increase of n and decrease of v_y and v_x.

5. Results and discussion

Quantity v_x - is a variation coefficient of the admissible limit value for the considered quality coefficient, it depends on: stability of level and range of impacts on a part as a whole; transmitting these outer impacts on the part and the given quality parameter; operational stability of this part; assembly quality of the part, etc. That is within the framework of the procedure of ion-plasma treatment, quantities v_x and m_x are not to be altered.

In this connection, depending on relation between v_y and v_x (σ_x and σ_y) it is possible to distinguish the following ranges. The first range is determined by the relation, the second range is determined by the relation $0.1\sigma_x \le \sigma_y$, and the third one is determined by $\sigma_y \ge 10\sigma_x$.

It is seen from relation (3), that in the second and the third ranges, rate increase $|U_\gamma|$ is great (efficient regulation) while raising safety factor from $n = 1$ up to $n \cong 1,5$. Then it substantially slows down. At the same time, influence of v_y on $|U_\gamma|$ complies with hyperbolic law, and it substantially influences on the quantity $|U_\gamma|$, especially in the range $\alpha \approx 1$.

Thus, depending on correlation between σ_y and σ_x, it is necessary to take different strategies on increase of faultless operation probability on the given quality parameter – to obtain increase of average value of the index, to obtain dispersion decrease of this index or to obtain both.

At the same time, it is necessary to keep in mind that increase of n requires searching of new materials and creating new technology as far as technology concern. As far as design concern, increase of n results in increase of structure mass, square of unsafe cross-sections, complexity of the structure itself, etc. Nevertheless, decrease of v_y or σ_y is connected with improvement of the technology process, decrease of coefficients of variations both properties of initial blanks and properties of the parameter at each production step including exclusion of manual labor and introducing automation. Moreover, to decrease dispersion, finishing technologies are especially important because they have positive technological heredity and low dispersion of output parameters, and vacuum ion-plasma technologies refer to them as shown in [2-4].

6. Evaluation technique of a technological procedure on permissible g-percentile alteration of quality coefficient

Thus, the use of suggested efficiency coefficient of technological procedures to increase working life and reliability of engineering parts objectively takes into account alteration of average value of the index, alteration of its dispersion. It also considers probability level of faultless operation at which this index is to be used; it reflects contemporary trends of reliability calculation of structures and parts, and shows that possibility of dispersion reduction of quality coefficients is a substantial reserve of increase of engineering parts reliability.

Acknowledgments

The reported study was funded by the Ministry of Science and Higher Education of the Russian Federation.

References

[1] Bogdanovich V I 2007 *Theoretical basis for ensuring the reliability of aircraft at the stage of their production* (Samara: Samara State Aerospace University Publishers)
[2] Barvinok V A, Bogdanovich V I, Dementiev S G 2014 *Modern technologies in aviation and rocket engineering* (Moscow: Mashinostroenie Publishers)
[3] Barvinok V A 2005 *Plasma in Technology: Reliability and Resource* (Moscow: Nauka i technologii Publishers)
[4] Barvinok V A 1990 *Stress control and properties of plasma coatings* (Moscow: Nauka i technologii Publishers)
[5] Barvinok V A, Bogdanovich V I 2008 *Technical physics* **53(1)** 64-68
[6] Gumen O, Bilyk I, Kruzhkova M 2020 *Lecture Notes in Civil Engineering* **47** 103-110
[7] Borysiuk J, Grabias A, Szczytko J, Bystrzejewski M, Twardowski A, Lange H 2008 *Carbon* **46(13)** 1693-1701
[8] Chen Z, Qiao L, Hillairet J, Song Y 2010 *Applied Surface Science* **492** 540-549
[9] Chang Y-Y, Cai M-C 2019 *Surface and Coatings Technology* **374** 1120-1127
[10] Potoczna-Petru D, Kępiński L, Krajczyk L 2005 *Materials Chemistry and Physics* **92(2-3)** 613-615

Ion-plasma coatings performance properties improvement obtained by arc deposition

V I Bogdanovich and M G Giorbelidze

Samara National Research University, 34, Moskovskoe Highway, Samara, 443086, Russian Federation

E-mail: bogdanovich@ssau.ru; m.giorbelidze@ssau.ru

Abstract. Influence of micro-droplet phase content in erosion products of cathodes made of various materials on coating properties produced by vacuum ion-plasma electro-arc method has been revealed. Amount of micro-droplet phase and its fractional content in a plasma jet have been researched. Coefficient standard distribution of micro-droplet phase content depending on cathode material and production conditions, distance from the cathode to the deposition surface, arc current and the temperature of the cathode surface have been obtained. General manufacturing methods and design concepts to reduce micro-droplet phase have been formulated.

1. Introduction

Development of coating deposition technologies supposes solving a number of tasks. They are the following: to choose deposited materials, to reach a certain property level, to model and to optimize the process, to designate deposition modes, etc. [1, 2]. All these aspects determine quality and product cost. One of these tasks is to control micro-droplet phase content (MDPh) contained in cathode erosion products. To solve this task, the researchers have applied the method of micro-droplet phase share determination in a vacuum chamber operating area and some manufacturing and design methods [1-2].

In case of obtaining nitrides, carbides and oxides, emergence of MDPh results in performance properties reduction of coatings (especially tribotechnical). However, in obtaining coatings of pure metals negative influence of MDPh is not practically observed, and in case of obtaining coatings made of multicomponent alloys (for instance, nickel-chrome-aluminium-yttrium), MDPh influences positively providing phase composition similarity of cathode and coating materials. The objective of this work is to generalize accumulated experience on MDPh reduction and development of general manufacturing principles of its control.

2. Experimental research of content of a micro-droplet phase

Experimental research of quantity and fractional content of MDPh was carried out by short-term putting of glass plates in a plasma jet. MDPh composition was researched using metallographic microscope, and quantitative assessment of a relative area occupied by MDPh was carried out using metallographic installation with an image analyzer in compliance with the relation:

Content from this work may be used under the terms of the Creative Commons Attribution 3.0 licence. Any further distribution of this work must maintain attribution to the author(s) and the title of the work, journal citation and DOI.
Published under licence by IOP Publishing Ltd

$$k_{cph} = \frac{S_{cph}}{S_0}$$

where S_{cph} – area occupied by micro-droplets, S_0 – total area of analyzed surface.

It should be noted, that absolute value of MDPh quantity could be determined by weighing only. However, it is impossible to separate MDPh from steam (atomic) phase. Measurement of the droplet phase coefficient k_{cph} allows to obtain only relative change of this magnitude in comparison with the certain reference value conditionally accepted as a unit. Analysis of these changes allows to choose events and modes reducing MDPh quantity, but obtained results do not allow to claim that exactly the same share of MDPh is in the coating.

During micro-droplet phase content measurements, it was considered that in initial period of the arcing, MDPh quantity was maximum because of uncontrolled compounds at the cathode surface. Under the following arcing, quantity of MDPh reduces at first, then it monotonically increases and in a certain time period t_0 it reaches stationary value in accordance with changes of mean integral temperature of the cathode surface. As a result, input of glass plates in a plasma jet is to be implemented after cathode running-in during the period $t \geq t_0 = \delta / a^2$, where δ - cathode thickness, a – temperature conductivity of cathode material. Hold up time of plates in a plasma jet was determined experimentally so that the background deposition of the coating would allow to carry out optical analysis for transparency of plates with MDPh.

Figures 1 and 2 show standard distributions of MDPh coefficient k_{cph} depending on cathode material and its manufacture conditions, distance from the cathode to the deposited surface, arc current, and the cathode surface temperature.

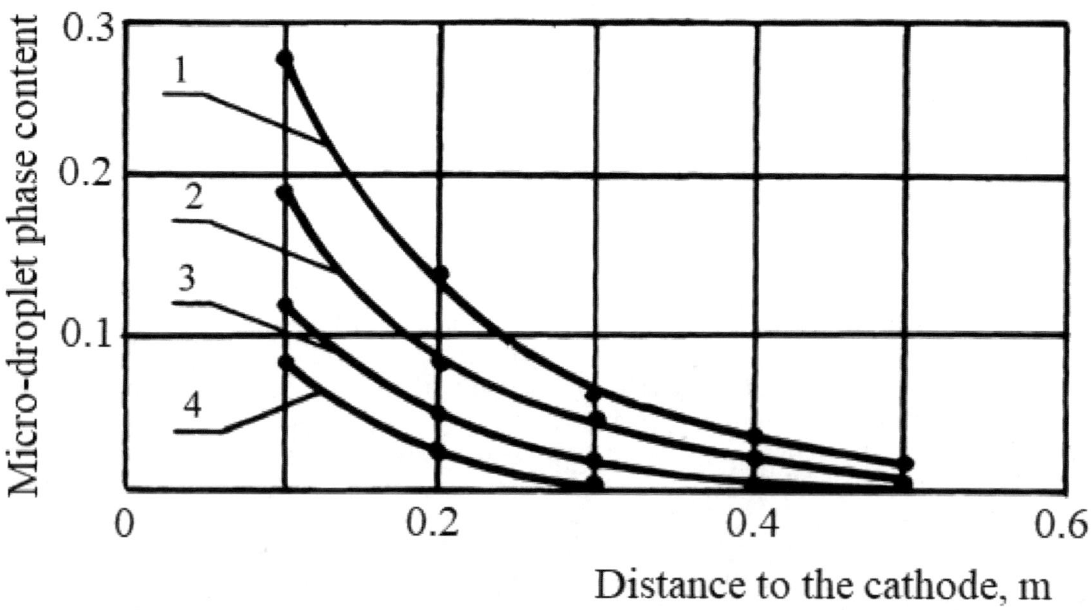

Figure 1. MDPh coefficient depending on the distance to the cathode for the materials: 1 - BT-6 (Ti-6Al-4V, Grade 5), 2 - BT-1-0 (Grade 2), 3 - BT-1-00 (ERTi-1), 4 - BT-1-00 (ERTi-1) vacuum annealing. Arc current 100 A.

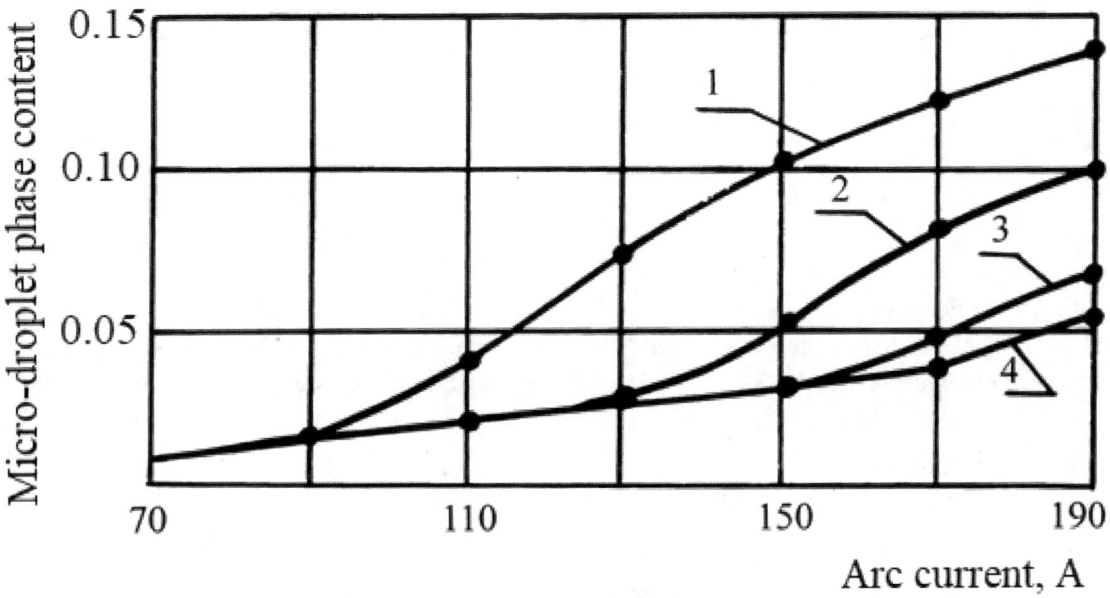

Figure 2. MDPh coefficient depending on arc current for various cathode thickness:

1 - δ=4.2 cm; 2 - δ=3 cm; 3 - δ=2.5 cm; 4 - δ=2 cm.

Research has shown that phase composition of cathode material and its manufacture method greatly influence on quantity and MDPh composition. For example, alloy BT-6 gives four times as much MDPh in comparison with pure titanium BT-1-00 (ERTi-1) subjected to vacuum annealing (Figure, points *1* and *4*). Besides, fraction MDPh composition from alloy BT-6 is substantially coarser. It is seen from Figure 1 that because of influence of gravitational separation MDPh quantity along the axis of a plasma jet in proportion to moving off from the cathode diminishes quickly enough due to going away of the coarsest fractions. Comparison of points *3* and *4* in Figure 1 allows evaluating influence of high-temperature annealing on MDPh content.

Scientific articles have more than once claimed that mean integral temperature of the cathode surface influences on MDPh quantity. To verify this statement, cathodes made of BT-1-00 with diameter $D = 8.0 \cdot 10^{-2}$ m and thicknesses δ: $4.2 \cdot 10^{-8}$ m; $3.0 \cdot 10^{-8}$ m; $2.5 \cdot 10^{-8}$ m; $2.0 \cdot 10^{-8}$ m were manufactured for the same metal plasma generator. Measurement results k_{cph} depending on arc current for the given cathode thicknesses are shown in Figure 2. Research has shown that monotone increase of k_{cph} for each thickness is observed. Moreover, if the arc current $I_{d,k}$, exceeds certain value, MDPh quantity increases greatly. It depends on delivered heat flow to the cathode and its cooling conditions.

Equation for mean integral temperature of the cathode surface is the following:

$$T_k = T_0 + \frac{0{,}3 I_d U_d \delta}{\pi R_k^2 \lambda}, \quad (1)$$

where U_d - voltage at the arc, δ - cathode thickness, R_k – cathode radius; λ - thermal conductivity of the material.

Having distinguished arc current values in Figure 2 at which we observe fast growth of MDPh for different thicknesses δ (point *1* - $I_d = 95$ A; point *2* - $I_d = 135$ A; point *3* - $I_d = 155$ A) and having substituted these values into (1), considering that $\lambda \cong 17$ W/m·K, $U_d = 22$ V, $T_0 = 323$ K, we get the following values of cathode temperatures T_k: 631 K, 636 K, 622 K, 585 K.

Similarity of all four values of temperature allows to state that characteristic increase of MDPh quantity in Figure 2 has the same thermal nature. When the mean integral temperature of the cathode surface reaches some characteristic value (in given case $T_{\kappa m} = 630$ K) MDPh quantity grows fast. Some reduction of $T_{\kappa m}$ for high currents (point 4 Figure 2) is apparently connected with the fact that at high currents, cathode spots are concentrated at the cathode center, and its surface temperature at the central part is to be determine according to the relations different from (1).

The same characteristic values have been found for other cathode materials too. That is why formula (1) is the following:

$$\frac{R_k^2}{\delta} \geq \frac{0{,}3 I_d U_d}{\pi\lambda\left(T_{\kappa m} - T_0\right)}, \tag{2}$$

where it is considered that $T_k \leq T_{km}$.

Equation (2) allows choosing dimensions of a cathode to achieve needed productivity or at the given cathode dimensions to find the value of a maximum arc current at which MDPh does not grow fast. For example, for the installation NNV-6.6-I1 with recommended dimensions of a titanium cathode ($R_k = 4\cdot 10^{-2}$ m, $\delta = 4.2\cdot 10^{-2}$ m) maximum arc current is not to exceed 95 A.

3. Results and discussion

The important aspect of MDPh quantity reduction is correct choice of initial cathode shape taking into account geometry of magnetic field lines. After long operation of a cathode, its surface takes the shape orthogonal to magnetic field lines (2). Later on, this stationary shape of the cathode surface is retained. At the same time, cathode surface shape depends not only on the direction of magnetic field lines, but also on the cathode material – diamagnetic, paramagnetic or ferromagnetic. Research has revealed that reaching this stationary shape is characterized with increased ejection of micro-droplet phase. As a result, when using new cathodes or at considerable changing of deposition mode (arc current or magnetic field induction) it is recommended to run in the cathode by means of deposition to the special shield to get stationary shape surface.

Emergence of such stationary cathode surface shape is getting clear at the basis of the following argumentation. Let at some initial moment, vector $\vec{B}$ is crossing the cathode surface at some angle that is it has perpendicular component B and parallel component $B_{\parallel}$ to the surface. Arc current with density $\vec{j}$ and its direction is perpendicular to the surface. Therefore, cathode spot rotation is appearing under the action of the force proportional to $\left[\vec{j}\cdot\vec{B}_{\parallel}\right]$ and as a result, azimuthal current $\vec{j}_{\varphi}$ is arising. Interaction of $\vec{j}_{\varphi}$ and $\vec{B}_{\perp}$ gives the force proportional to $\left[\vec{j}_{\varphi}\cdot\vec{B}_{\perp}\right]$, this force is directed to the cathode center (under redirection of $\vec{B}_{\perp}$ directed from the cathode center). As a result, cathode spots are in the cathode center for a longer time providing removal of large amount of material at this place. In the course of such selective cathode making, value $B_{\parallel}\to 0$ and cathode spots rotation with preferred staying in the center is turning into chaotic motion in this area with uniform surface development (self-regulation).

Under increasing of magnetic field value perpendicular to the cathode surface, rotation rate of the cathode spot over the surface is also increasing. However, starting with certain values $\vec{B}$ cathode spot is squeezed in the central part of the cathode. It is deduced from experiments that at a certain value of magnetic field induction, cathode spot localization in the central part of the cathode results in the temperature of the cathode surface at this aria exceeds critical value and MDPh content in a plasma jet increases sharply. Value of magnetic field induction at which MDPh output increases, depends on some factors including a specific design of a vacuum arc generator and a solenoid developing outer magnetic field.

4. Conclusions

Executed researches allow to formulate general manufacturing and design principles of MDPh reduction:

1) to choose cathode material purposefully and to prepare it specially before deposition;
2) it is initially necessary to give cathodes the shape that would remain under its erosion;

3) to optimize deposition modes to minimize MDPh quantity;

4) when developing new plasma generators, it is necessary to optimize electrodes geometry, electric and magnetic field lines configuration.

Acknowledgments

The reported study was funded by the Ministry of Science and Higher Education of the Russian Federation.

References

[1] Barvinok V A 2005 *Plasma in Technology: Reliability and Resource* (Moscow: Nauka i technologii Publishers)

[2] Barvinok V A 1990 *Stress control and properties of plasma coatings* (Moscow: Nauka i technologii Publishers)

[3] Barvinok V A, Bogdanovich V I 1999 *Physical basis and mathematical modeling of vacuum ion-plasma deposition* (Moscow: Nauka i technologii Publishers)

[4] Valleti K, Jyothirmayi A, Ramakrishna M, Joshi S V 2020 *Journal of Vacuum Science and Technology A: Vacuum, Surfaces and Films* **29(5)** 051515

[5] Zhang J, Guo W, Zhang Y, Guo Q, Wang C, Zhang L 2009 *Thin Solid Films* **517(17)** 4830-4834

[6] Borysiuk J, Grabias A, Szczytko J, Bystrzejewski M, Twardowski A, Lange H 2008 *Carbon* **46(13)** 1693-1701

[7] Gumen O, Bilyk I, Kruzhkova M 2020 *Lecture Notes in Civil Engineering* **47** 103-110

Analysis of contact strength of spherical roller transmission with double-row pinion

M E Lustenkov and A N Moiseenko
Belarusian-Russian University, Mira Avenue 43, Mogilev, 212000, Republic of Belarus

E-mail: lustenkov@gmail.com

Abstract. This paper presents a detailed description of the design of the spherical roller transmission. The pinion has two rows of rollers; the axes of the rollers are located on conical surfaces. When the transmission is operating, the pinion performs a regular precession and the rollers are in contact with the teeth of the stationary and driven central gears. The transmission provides high reduction ratios in a lightweight, compact package. The center of mass of the pinion is located on the axis of rotation of the drive shaft, which reduces the axial moment of inertia and improves the balance of the system compared to double-pinion planetary gears mounted on the drive shaft with eccentricity. Rolling bodies are used in place of teeth, which leads to improved efficiency. The main performance criterion is the contact strength of the teeth of the driven central gear. An equation to calculate maximum contact stresses has been derived by using Hertz theory. An algorithm for calculating the number of the rollers simultaneously engaged and transferring the load is proposed.

1. Introduction

Power drives with high-speed motors and slowly rotating drive shafts find application in various fields: positioning and servo systems, kinematic chains of robots, the aerospace industry. Besides, they are widely used in household appliances. To achieve high reduction ratios in drives with a compact design, mechanical transmissions of various types are used. Simple planetary gears can produce reduction ratios up to 50 ... 60. However, their reliability and efficiency significantly depend on the way in which rotation is transmitted from the pinion to the driven shaft [1]. Due to the use of double-pinion planetary gears, the range of reduction ratios can be increased significantly; the disadvantages are increased slip and low efficiency [2]. Power losses can be reduced by replacing teeth with rolling bodies [3]. Transmissions with the pinion that does not move in a plane but performs a regular precession are referred to as nutation or precessional transmissions [4–7]. They have a lower total axial moment of inertia and smaller radial dimensions. Since the pinion performs a spherical motion, the transmissions fall within the category of spherical mechanisms, which can be used in various fields [8, 9]. To increase the number of rollers simultaneously transferring the load, a number of designs of spherical roller transmissions (SRT) with rows of rollers arranged coaxially have been developed [10]. To reduce radial dimensions of the transmission, the rows of rollers in most precessional transmissions are arranged axially relative to each other [11]. This paper presents a description of a new SRT design developed to increase the transmitted unit load. The study is aimed to develop a new transmission design and estimate the strength of the roller engagement.

Content from this work may be used under the terms of the Creative Commons Attribution 3.0 licence. Any further distribution of this work must maintain attribution to the author(s) and the title of the work, journal citation and DOI.
Published under licence by IOP Publishing Ltd

2. SRT model and principle of operation

Figure 1 shows the main elements of the SRT 3D model created in NX and intended for subsequent kinematic and strength calculations. The rollers are combined with a pinion and are made as cylindrical teeth.

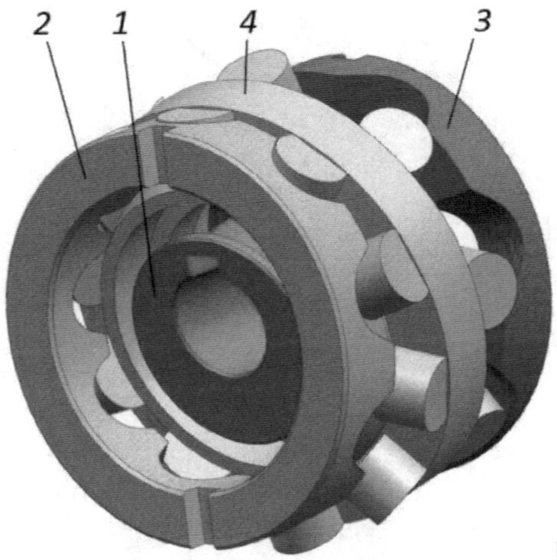

Figure 1. SRT model (main elements): 1 – eccentric; 2 – driven central gear; 3 – stationary central gear; 4 – pinion.

The drive shaft of SRT has a section inclined to its axis at a nutation angle Θ. The inclination is provided by means of eccentric 1 installed on the drive shaft. Pinion 4 is mounted on the eccentric by means of bearings. The rollers are arranged in two rows with uniform angular spacing; their axes in each row are located on the conical surface and are symmetrical with respect to the axis of rotation of the pinion. One row of the rollers is in contact with the teeth of stationary central gear 3, which is fixed in the housing, and the second row of the rollers engages with the teeth of central gear 2, which is connected to the driven shaft. The teeth of the central gears have a constant height. The rollers, alternately getting into the extreme left and right positions where they extend into tooth spaces of the gears to the maximum depth, occupy positions where their axes are perpendicular to the axes of the transmission. In this way, the maximum length of the contact lines and the transmitted unit load are reached. Due to the fact that the rollers can rotate about their axes, sliding in contact is partially replaced by rolling, which leads to increased efficiency of the transmission.

Equations for determining the reduction ratio for SRT with a double-row pinion are similar to those for double-pinion planetary gears. The number of rollers in rows performs the function of the number of gear teeth. The range of produced reduction ratios can be estimated as 16 ... 200 for transmissions with diametrical dimensions not exceeding 250 mm.

The main reliability criterion for SRT is the contact strength of the transmission parts [12]. To perform the check calculation, the following data must be taken into account: the nominal torque on the driven shaft T_2, N·m, materials used to make parts and heat treatment techniques, the reduction ratio, the number of teeth on the driven gear Z_2 and the stationary gear Z_3, respectively, the number of rollers in the pinion rows n_{s2} and n_{s3}, respectively, as well as geometric parameters of the transmission.

3. SRT contact strength analysis

Let us consider the interaction of one row of rollers with the driven central gear, as it is subjected to higher loads. The middle section is taken as the section for strength calculations, the parameters are assigned the subscript m (figure 2). Space curve 4 (center curve) is the trajectory of point C lying on the axis of roller 1 when it moves relative to the driven central gear.

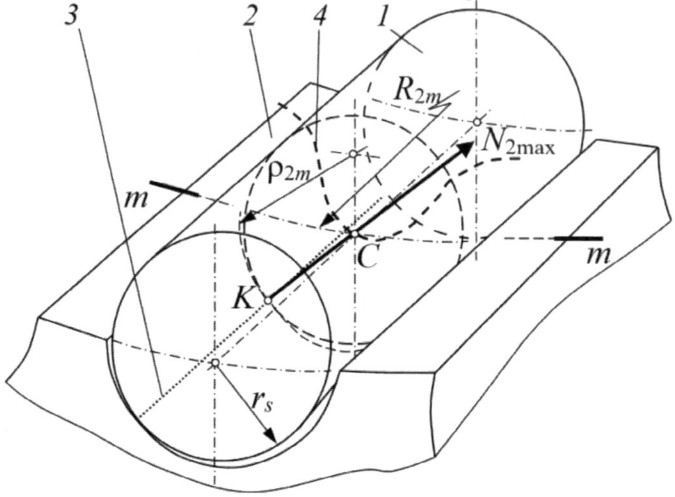

Figure 2. Schematic representation of roller contact with central gear teeth: 1 – roller; 2 – toothed profile; 3 – contact line; 4 – center curve in the middle section.

Maximum contact stresses $\sigma_{H\max}$ are calculated using the formula modified for interaction of a cylinder and a concave surface [13] according to Hertz theory

$$\sigma_{H\max} = K_{Gs} \cdot \sqrt{N_{2\max} \cdot \frac{\rho_{2m} - r_s}{\rho_{2m} \cdot r_s}}, \quad (1)$$

where K_{Gs} is the coefficient depending on the properties of the materials of contacting parts, MPa$^{1/2}$; $N_{2\max}$ is the maximum value of the normal force in the contact between the roller and the teeth, N; ρ_{2m} and r_s are the radii of curvature of the contacting surfaces (the toothed profile in the middle section and the roller, respectively), mm.

The coefficient K_{Gs} is calculated using the formula

$$K_{Gs} = 0.564 \cdot (\theta_1 + \theta_2)^{-0.5}, \quad (2)$$

where θ_1 and θ_2 are the coefficients depending on the properties of materials of contacting bodies.

For steel parts K_{Gs} = 473.6 MPa$^{1/2}$. Then

$$\theta_{1(2)} = \left(1 - \mu_{1(2)}^2\right) \cdot E_{1(2)}^{-1}, \quad (3)$$

where $\mu_{1(2)}$ is the Poisson's ratio of the material of roller (1) and the teeth of driven central gear (2), respectively; $E_{1(2)}$ is the elastic modulus. For steel parts $E_1 = E_2 = 2.1 \cdot 10^5$ MPa, $\mu_1 = \mu_2 = 0.3$.

To calculate the radius of curvature, let us consider the scheme of the middle section of the toothed profile in figure 3. We approximate the complex profile of surface 1 replacing it with circular arc 2. From geometric constructions, using the properties of the chords of the circle, we obtain

$$\rho_{2m} = 2 \cdot R_m^2 \cdot \sin^2(\psi_{z0}) \cdot r_s^{-1} + 0.5 \cdot r_s, \quad (4)$$

where R_m is the radius of the basic sphere with the center curve in the middle section, mm; ψ_{z0} is the central angle, rad, calculated using the formula

$$\psi_{z0} = \frac{\pi \cdot r_s}{R_m \cdot Z_2 \cdot \sin(\alpha_{2m})}, \quad (5)$$

where α_{2m} is the mean value of the slope of the center curve determined in the middle section of the driven central gear.

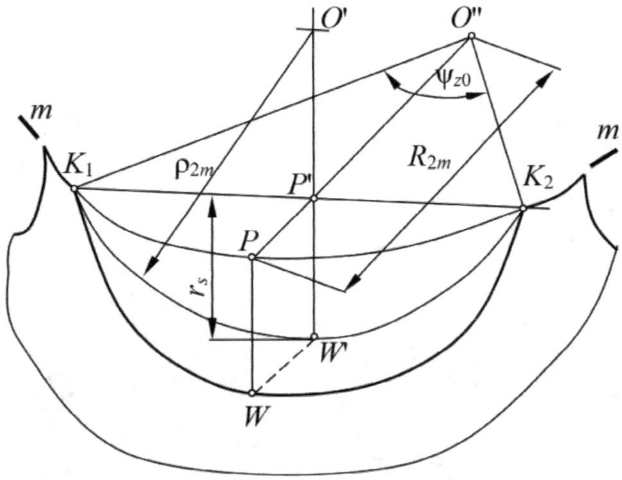

Figure 3. Scheme for determining radius of toothed profile curvature in middle section.

To calculate the slope of the center curve, we simplify the problem by replacing it with a piecewise helix and studying its unfolded view. The resulting curve with constant values of slopes is a set of straight line segments; the slopes are calculated according to the formula

$$\alpha_{2m} = \text{arctg}\left(2 \cdot Z_2 \cdot \Theta \cdot \pi^{-1}\right), \quad (6)$$

where R_{2m} is the radius of the circle with the point K of application of the normal force, mm.

The maximum normal load on one roller, N, is calculated as follows

$$N_{2\max} = \frac{T_2 \cdot 10^3}{R_{2m} \cdot n_{s2} \cdot K_{p2} \cdot K_n \cdot \sin\left(\alpha_{2m}\right)}, \quad (7)$$

where n_{s2} is the number of the rollers in the pinion row interacting with the driven central gear; K_{p2} is the coefficient that takes into account the number of the rollers simultaneously transferring the load; K_n is the coefficient that takes into account uneven load distribution through the flows.

The scheme for determining the coefficient K_{p2} (figure 4, a) shows the roller in the leftmost position with the axis in a vertical position. As the transmission continues to run, the roller starts receding from the contact with the toothed profile of the flat gear, and the lowest leftmost point B of the roller is the last to move out of engagement. By connecting these points for all the rollers in the row, we obtain circle 2 (figure 4, b). Central gear 4 is bounded on the right by plane 1. Let us suppose that the rollers with the points which lie furthest to the left of plane 1, i.e., "penetrating" into the toothed profile, engage with the gear, and the part of the circle to the right of this plane unites the leftmost points of the rollers that are not in contact with the gear teeth.

Then the coefficient K_{p2}, which is determined as the number of the rollers simultaneously transferring the load relative to the total number of the rollers, can be calculated as the ratio of the angle φ_r to the angle $2{\cdot}\pi$. The angle φ_r is the operating angle and the angle equal to $2{\cdot}\pi - \varphi_r$ is the idling angle.

To find the operating angle, let us first calculate the angle φ_s using the parametric equation of circle 1derived earlier, which describes the change in the z-coordinate depending on the parameter φ, assuming that in the equation $z(\varphi) = L_s$ (figure 4, b). After transformations we obtain

$$\sin\left(\varphi_s\right) = \frac{L_s - \left(L_{k\max} + r_s \cdot \cos(\Theta)\right) \cdot \cos(\Theta)}{\sin(\Theta) \cdot \sqrt{R_{2\min}^2 + \left(L_s + r_s\right)^2 - \left(L_{k\max} + r_s \cdot \cos(\Theta)\right)^2 \cdot \cos^2(\Theta)}}, \quad (8)$$

where $L_{k\max}$ and $R_{2\min}$ are the geometric parameters shown in figure 4, a.

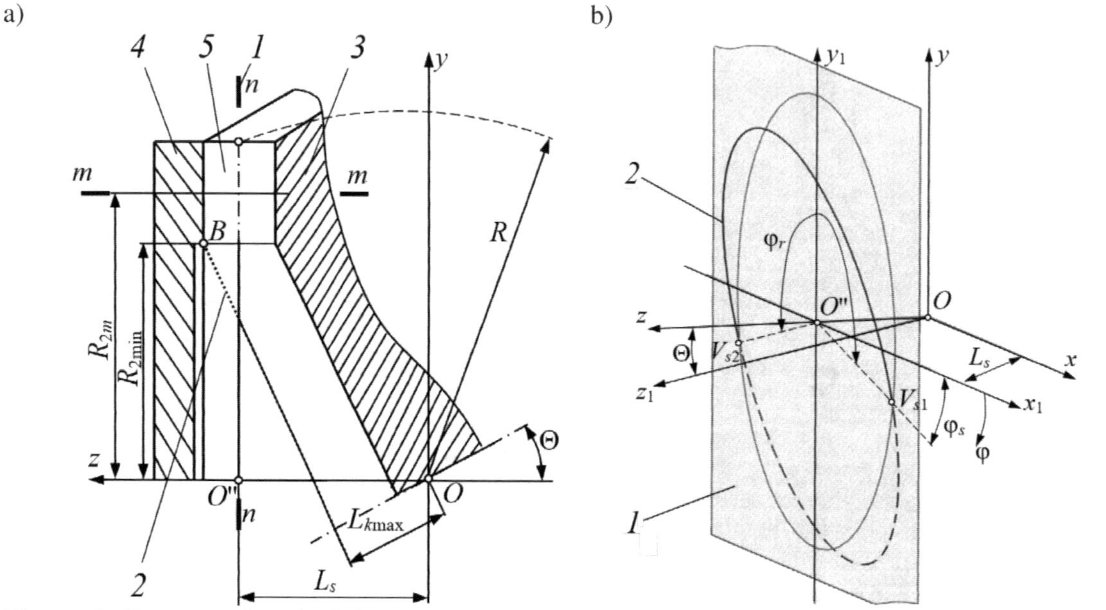

Figure 4. Contact ratio calculation: a – schematic representation of the contact; b – computational scheme; 1 – plane bounding the toothed profile; 2 – circle with the lowest leftmost points of rollers; 3 – pinion; 4 – driven central gear; 5 – roller.

Finally, the contact strength condition is written as follows:

$$\sigma_{H\max} = K_{Gs} \cdot \sqrt{\frac{T_2 \cdot 10^3}{R_{2m} \cdot n_{s2} \cdot K_{p2} \cdot K_n \cdot \sin(\alpha_{2m})} \cdot \frac{\rho_{2m} - r_s}{\rho_{2m} \cdot r_s}} \leq [\sigma_H], \tag{9}$$

where $[\sigma_H]$ are the allowable contact stresses, MPa.

Using the above formulas, a check calculation for the transmission with a reduction ratio of 36 and the number of teeth and rollers of $Z_2 = 9$, $Z_3 = 7$, $n_{s2} = 10$, $n_{s3} = 8$ was performed. The following parameters were used in the calculation: the torque on the driven shaft $T_2 = 50$ N·m, the radius of the basic sphere determining the dimensions of the transmission $R = 30$ mm (figure 4, a), $r_s = 5$ mm and $\Theta = 0.2$ rad. The transmission parts were made of steel with accuracy grade 7; the experimental coefficient was taken to be $K_n = 0.9$. According to the results of the calculations performed using Mathcad, the maximum values of contact stress were determined to be $\sigma_{H\max} = 605$ MPa. The coefficient taking into account the proportion of the rollers engaged simultaneously was found to be $K_{p2} = 0.538$.

4. Conclusions

On the basis of Hertz theory, the equation which can be used to calculate maximum contact stresses in the teeth of the SRT central driven gear has been derived. Allowable stresses, equal to or greater than the calculated values, can be obtained by using the most common and relatively low-cost types of alloy steels and appropriate heat treatment techniques. The algorithm for calculating the number of rollers simultaneously transferring the load to SRT has been developed. The calculations performed using this algorithm have shown that more than half of the rollers of the transmission with the above parameters are simultaneously engaged and their ratio to the total number of the rollers does not depend on the reduction ratio. The analysis of the developed model shows that with the same geometric parameters of the transmission specified above, the maximum contact stresses decrease as the reduction ratio increases, which is due to the greater number of teeth on the central gears and rollers. However, the number of teeth does not exceed some limiting value subject to the condition $\rho_{2m} \geq r_s$.

References

[1] Chen, E. and Walton D 1990 The optimum design of KHV planetary gears with small tooth differences *International Journal of Machine Tools and Manufacture* **30**(1) 99–109

[2] Liao Y G and Chen M-Y 2017 Analysis of multi-speed transmission and electrically continuous variable transmission using lever analogy method for speed ratio determination *Advances in Mechanical Engineering* **9**(8) 1–12

[3] Efremenkov E A and Bonnard E 2020 Power Parameters Automated Calculation for Transmission with Intermediate Rolling Bodies and Free Cage *IOP Conf. Ser.: Mater. Sci. Eng.* **795** 012025

[4] Molyneux W G 1997 The internal bevel gear and its applications *Proc. Inst. Mech. Eng.* G **211** 39–61

[5] Hong J, Yao L, Ji W and Huang Z 2015 Kinematic modeling for the Nutation Drive Based on Screw Theory *Procedia CIRP* **36** 123–128

[6] Wang G X and Guan T M 2009 Modeling of Nutation Drive with Rolling Teeth *Applied Mechanics and Materials* **16–19** 708–712

[7] Frumusanu G R and Oancea N 2013 Technological solution to profile and generate the teeth of central gear for precessional gear drives *Int. J. Adv. Manuf. Technol* **67** 687–699

[8] Bai S and Angeles J 2009 The Design of Spherical Multilobe-cam Mechanisms *Proc. IMechE* vol 223 part C: *J. Mechanical Engineering Science* 473–482

[9] Zhang L, Pan C, Xu X, Xu H and Zhang Z 2016 An investigation of cam-roller mechanism applied in sphere cam engine Journal of Central South University **23**(4) pp 825–833

[10] Lustenkov M E and Lustenkova E S 2020 Load Capacity of Spherical Roller Transmission with Double-Row Pinion Load Capacity of Spherical Roller Transmission with Double-Row Pinion *IOP Conf. Ser.: Mater. Sci. Eng.* **795** 012020

[11] Bostan I, Dulgheru V and Ciobanu R 2016 Elaboration and research of planetary precessional multiplier type K-H-V *IOP Conf. Ser.: Mater. Sci. Eng.* **147** 012002

[12] Lustenkov M E 2010 Planetary Ball Transmissions: Strength Calculations *Russian Engineering Research* **30**(9) 862–866

[13] Birger I A, Shorr B F and Iosilevich G B 1993 *Strength calculation of machine parts* (Moscow)

IOP Conf. Series: Materials Science and Engineering 1118 (2021) 012007 doi:10.1088/1757-899X/1118/1/012007

Experimental estimation of efficiency and kinematic accuracy of a spherical roller transmission

E S Lustenkova
Belarusian-Russian University, Mira Avenue 43, Mogilev, 212000, Republic of Belarus

E-mail: fittsova@gmail.com

Abstract. Spherical roller transmissions with double-row pinions provide high reduction ratios in drives with a compact and lightweight design. The transmissions can be used as mechanical modules in kinematic chains of robots and manipulators. The pinion has two rows of coaxially arranged rollers and, when in operation, performs a regular precession. The paper examines the characteristics of an experimental model of a speed reducer with a spherical roller transmission with a reduction ratio of 44. The arrangement of the test rig as well as the test instruments and methods are described. The results of experimental study determining the efficiency of a spherical roller transmission depending on the torque being transmitted and the rotational speed of the drive shaft are presented. It has been found that the mean efficiency of the speed reducer is within the range of 0.7 ... 0.8 and its maximum value is 0.81, which corresponds to the efficiency of worm gears with similar characteristics. The theoretical calculations for determining the efficiency, which were obtained in previous studies, have been confirmed; the value of the reduced friction coefficient has been calculated. The constancy of the efficiency and the reduction ratio of the experimental model of the speed reducer and the factors influencing this constancy have been investigated.

1. Introduction

Mechanical transmissions with high reduction ratios are used in drives of aerospace equipment, robots and manipulators, pipe valves, etc. When designing a low-speed drive, a set of cylindrical and bevel gears connected in cascade can be used; other types of gears that can achieve high reduction ratios in one stage can be employed as well. The second option is more promising, since the cascade connection of gears leads to larger dimensions and greater weight of the drive as well as an increase in backlash; the rigidity, accuracy and reliability of the system are reduced. Due to the use of compact coaxial transmissions providing high reduction ratios in one stage, the actuator is brought closer to the engine, the weight and dimensions of the drive are reduced. To achieve these goals, various types of mechanisms are used, and cycloidal pin planetary gears are the most promising ones [1]. High reduction ratios can be achieved in schemes of transmissions with double-row pinions. However, as slip increases, efficiency tends to decrease; the pinion mounted on the drive shaft with eccentricity needs to be balanced.

Spherical mechanisms are compact in size, provide additional kinematic capabilities [2, 3] and are used to produce transmission mechanisms. Due to the use of rolling elements instead of teeth, sliding is replaced with rolling, which results in increased efficiency. Spherical transmissions with centers of mass (axes) of rolling bodies not fixed on the pinion [4] are characterized by excessive noise, wear and

Content from this work may be used under the terms of the Creative Commons Attribution 3.0 licence. Any further distribution of this work must maintain attribution to the author(s) and the title of the work, journal citation and DOI.
Published under licence by IOP Publishing Ltd

potential jamming. Motion transmission in precessional (nutation) transmissions is based on the phenomenon of regular precession [5–7]. These transmissions meet specified requirements in many respects and can be regarded as spherical mechanisms, since trajectories of the pinion points lie on spherical surfaces. A spherical roller transmission (SRT) with a pinion with two rows of rollers mounted coaxially [8] can produce reduction ratios in the range of 16 ... 200 and has a number of advantages, such as improved balance and higher load capacity.

This paper examines experimentally a model of the developed SRT in order to determine the instantaneous and mean efficiency of SRT as well as the constancy of the instantaneous reduction ratio depending on the rotational speed of shafts and the transmitted torque.

2. Object, instrumentation and methods used for SRT testing

In this study, an experimental model of a speed reducer with a double-row pinion SRT and with graphite grease applied during assembly was chosen as the test object. One row of SRT rollers interacts with a periodic closed groove formed by outer cams mounted in the housing; the second row of rollers interacts with a groove made on the inner cam connected to the driven shaft. In contrast to available gear transmission mechanisms, the load capacity is increased due to a larger number of parallel power flows. All the rollers of the row are simultaneously in contact with the outer cams, and about half of them are in contact with the inner cam. Sliding in the transmission is partially replaced by rolling, since rollers have the ability to rotate around their axes being in contact with grooves. The center of mass of the pinion is located on the transmission axis and coincides with the fixed point of spherical motion, which improves the balance of the mechanism and reduces the starting torque. The reducer has a reduction ratio of $i_{12}^{(3)} = -44$, which is determined by the number of periods of grooves formed by outer cams – $Z_2 = 11$, by inner cams – $Z_3 = 15$ and the number of the rollers of the outer row n_{s3} and the inner row n_{s2}, which are one more than the numbers of periods of the corresponding grooves. The maximum diameter of the reducer housing is 82 mm.

The test rig (figure 1) operation is based on the open power flow method. The reducer is placed in the kinematic chain of the rig between the torque sensor and the rotational speed sensor.

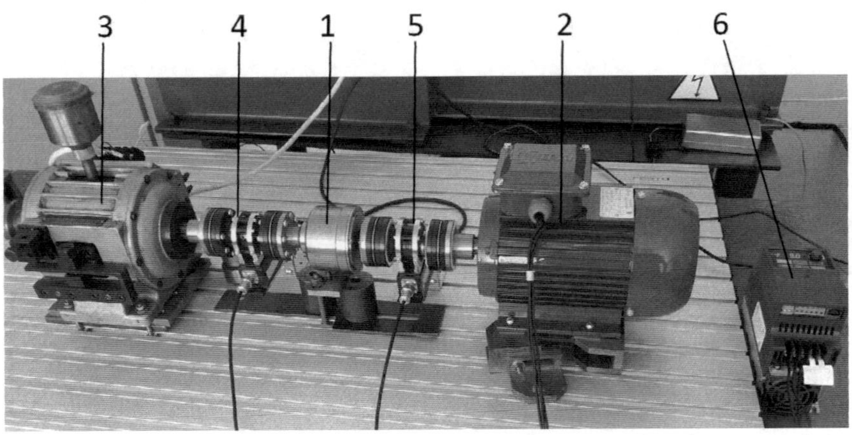

Figure 1. Test rig arrangement: 1 - speed reducer; 2 - asynchronous electric motor; 3 - powder brake; 4, 5 - torque sensor and rotational speed sensor; 6 - frequency converter.

Torque measurement is based on deformation of strain gauge elements and determination of the magnitude of unbalance of the strain gauge bridge circuit. Rotational speed measurement involves recording pulses by means of the infrared detector on the rotor, which are generated from the infrared emitter mounted on the stator.

A number of well-known methods were used in the design of the experiment [9]. Instantaneous efficiency was calculated as the ratio of the powers on the input and output shafts

$$\eta = \frac{T_2 \cdot n_2}{T_1 \cdot n_1}, \tag{1}$$

where T_1, T_2 are torques on the drive and driven shafts, respectively; n_1 and n_2 are rotational speeds of the drive and driven shafts, respectively.

The kinematic accuracy of SRT is characterized by the error of the actual instantaneous reduction ratio (its deviation from its nominal value). It was defined as the difference between the nominal $i_{12}^{(3)}$ and real values.

$$\Delta i = \left|i_{12}^{(3)}\right| - \left|\frac{n_1}{n_2}\right|. \tag{2}$$

To perform spectral analysis based on the discrete Fourier transform, the data were centered. For instantaneous efficiency, the centered values were determined according to the formula

$$F\eta_j = \eta_m - \frac{1}{N_r} \cdot \sum_{j=1}^{N_r} \eta_j \,. \tag{3}$$

where N_r is the number of measurements made during one revolution of the driven shaft, η_m is the mean value of the efficiency during a specified time interval.

Fourier coefficients were calculated according to the following formulas:

$$C_{1k} = \frac{2}{N_r} \cdot \sum_{j=1}^{N_r} F\eta_j \cdot \cos\left(\frac{2 \cdot \pi \cdot k \cdot j}{N_r}\right);\ C_{2k} = \frac{2}{N_r} \cdot \sum_{j=1}^{N_r} F\eta_j \cdot \sin\left(\frac{2 \cdot \pi \cdot k \cdot j}{N_r}\right), \tag{4}$$

where $j = 1 \ldots N_r$, $k = 1 \ldots N_r/2$, since the maximum number of spectral components cannot be greater than half of the measurements.

For each k-th harmonic, the amplitude was determined as follows

$$A_k = \sqrt{C_{1k}^2 + C_{2k}^2}\,. \tag{5}$$

Similar calculations were performed for the reduction ratio deviation Δi.

3. Test results and their analysis

The values of the mean efficiency at different torques on the driven shaft and different rotational speeds are given in table 1.

Table 1. Results of experimental study of SRT efficiency.

T_2, N·m	η_m	T_2, N·m	η_m	T_2, N·m	η_m	T_2, N·m	η_m	T_2, N·m	η_m
$n_1 = 500$ min^{-1}		$n_1 = 750$ min^{-1}		$n_1 = 1000$ min^{-1}		$n_1 = 1250$ min^{-1}		$n_1 = 1500$ min^{-1}	
3	0.689	3	0.678	3	0.745	3	0.687	2	0.669
5	0.708	5	0.713	5	0.79	6	0.726	5	0.732
8	0.725	8	0.735	8	0.8	8	0.741	10	0.749
12	0.724	12	0.735	10	0.81	10	0.744	12	0.676
-	-	-	-	12	0.775	-	-	-	-

Note: the table shows the mean values of rotational speeds n_1 and torques T_2

Figure 2 shows the results of studies with addition of fifth-order polynomial trendlines.

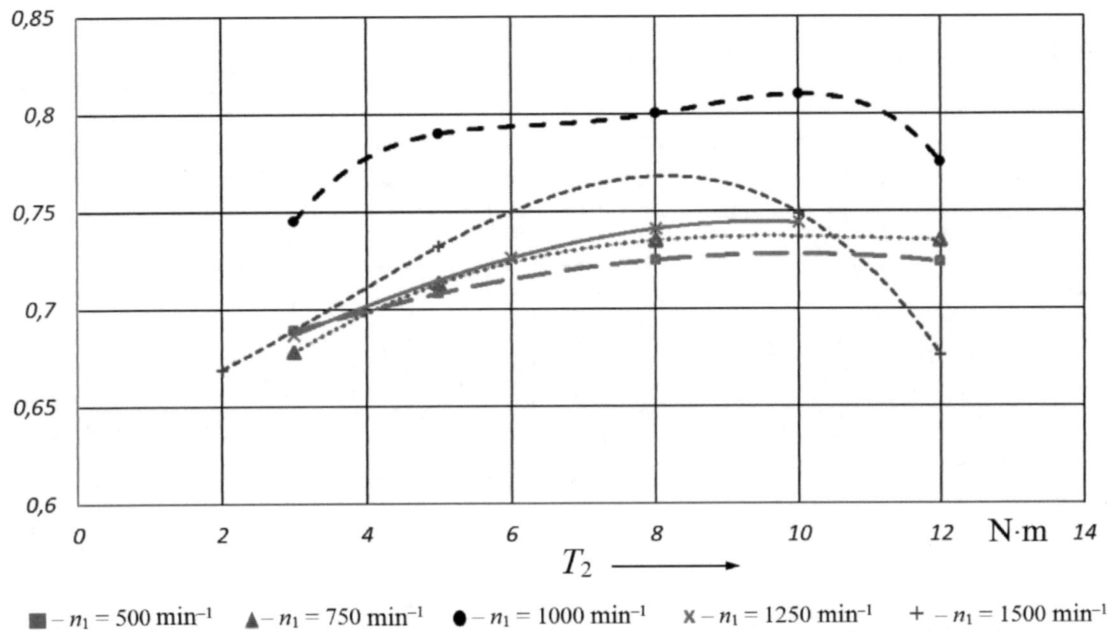

Figure 2. Graphs of SRT efficiency change

As the load increases, the efficiency of the transmission increases as well reaching a certain maximum. Then a decrease is observed, which indicates the existence of optimal values of the transmitted torque according to the criterion of maximum efficiency. Besides, there is an optimal rotational speed; a decrease or increase in its values leads to higher power losses. The graph in figure 3 shows the change in the instantaneous efficiency during one revolution of the driven shaft for T_2 = 10 N·m and n_1 = 1000 min^{-1}.

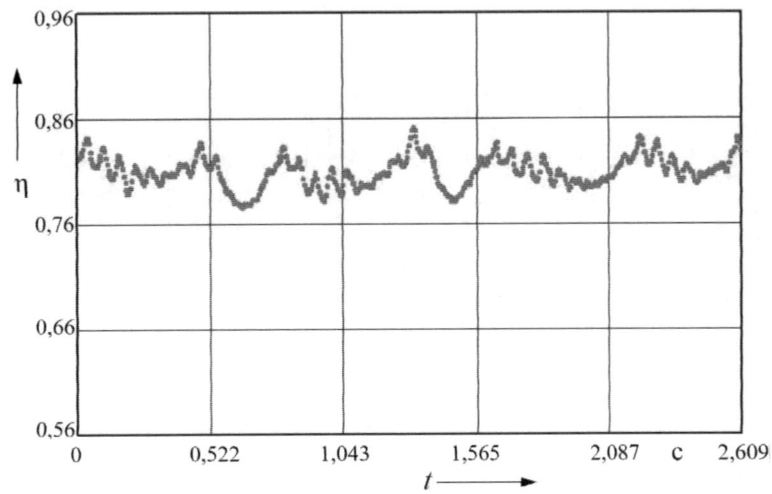

Figure 3. Measurements characterizing change in instantaneous SRT efficiency.

When the geometric parameters of SRT, the reduced coefficient of friction between the rollers and grooves f = 0.02 and the coefficient of friction in the bearings of the pinion f_b = 0.003 were substituted into the theoretically obtained SRT model, developed to determine the mean efficiency [10], the calculated efficiency was η = 0.762, which corresponds to the values obtained during the experiments.

However, the efficiency in the theoretical model does not depend on the load and rotational speeds of the shafts. The coefficient *f* is reduced and takes into account sliding and rolling.

From the spectrum in figure 4 we can see that the sixth harmonic has the highest amplitude, which corresponds to half the number of rollers in the inner row ($n_{s2} = 12$). Thus, the constancy of the efficiency of the transmission under study is affected by uneven spacing between the rollers in the inner row. The forty-first harmonic is an indication of the influence of inaccuracies in manufacture of the drive shaft with the pinion on the eccentric and mounting errors.

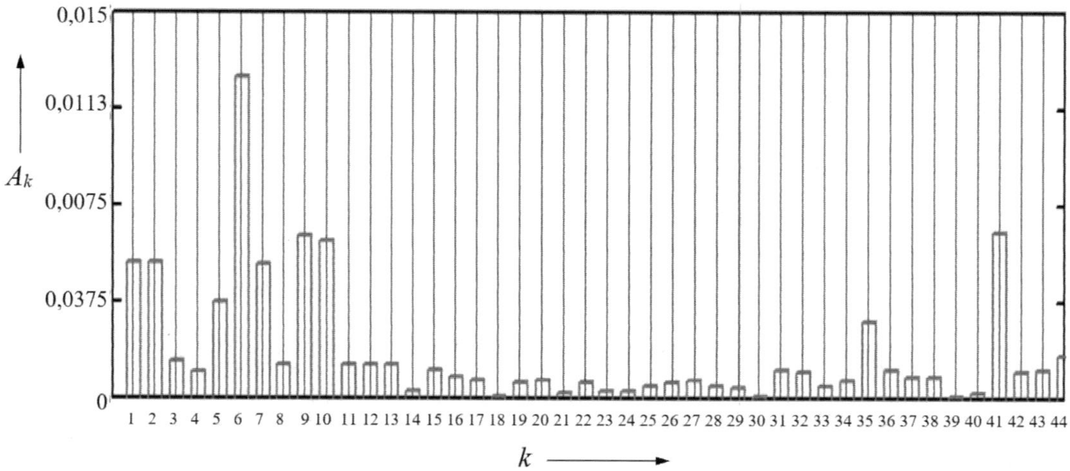

Figure 4. Data spectrum of instantaneous efficiency measurement for SRT experimental model.

The kinematic error (constancy of the reduction ratio) is most affected by the low-frequency components of the spectrum, which indicates the presence of errors in the manufacture of the driven shaft and driven cam and their incorrect mounting (figure 5).

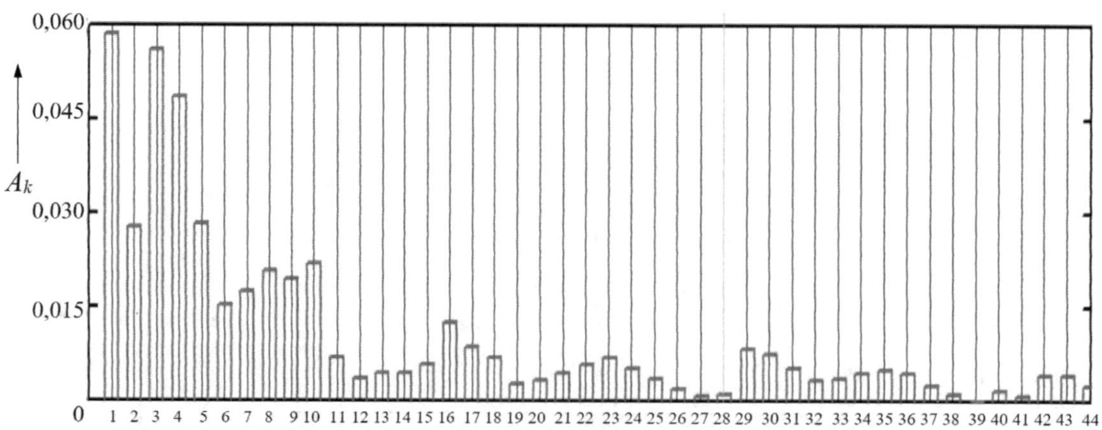

Figure 5. Data spectrum of kinematic accuracy calculations for SRT experimental model.

The error in the instantaneous reduction ratio can be considered insignificant, since its maximum values do not exceed 0.14% of the nominal value, while the instantaneous efficiency deviations from the mean values are within the range of 1.6 ... 1.8%.

4. Conclusions

The results of tests of experimental models of speed reducers and motor-reducers with SRT: theoretical calculations for estimating the mean efficiency have been confirmed, a mean efficiency of

0.7 ... 0.8 has been achieved with a reduction ratio of 44, which corresponds to the efficiency of worm gears with similar reduction ratios. The optimal rotational speeds of the shafts and the transmitted torque have been determined according to the criterion of minimum power losses in the engagement.

The results of the experimental data processing have demonstrated that the constancy of the efficiency of the transmission under study is most affected by uneven spacing between the rollers of the inner row as well as inaccuracies in the manufacture of the drive shaft and mounting errors. The kinematic error (the reduction ratio stability) of the experimental model of the transmission is mainly influenced by the error in the manufacture of the driven shaft and its incorrect mounting.

References

[1] Efremenkov E A and Bonnard E 2020 Power Parameters Automated Calculation for Transmission with Intermediate Rolling Bodies and Free Cage *IOP Conf. Ser.: Mater. Sci. Eng.* **795** 012025

[2] Bai S and Angeles J 2009 The design of spherical multilobe-cam mechanisms *Mechanical Engineering Science* **223** 473–482

[3] Zhang L, Pan C, Xu X, Xu H and Z. & Zhang 2016 An investigation of cam-roller mechanism applied in sphere cam engine *Journal of Central South University* **23**(4) 825–833

[4] Wang G X and Guan T M 2009 Modeling of Nutation Drive with Rolling Teeth *Applied Mechanics and Materials* **16–19** 708–712

[5] Molyneux W G 1997 The internal bevel gear and its applications *Proceedings of the Institution of Mechanical Engineers* G: *Journal of Aerospace Engineering* **211**(1) pp 39–61

[6] Hong J, Yao L, Ji W and Huang Z 2015 Kinematic Modeling for the Nutation Drive Based on Screw Theory *Procedia CIRP* **36** 123–128

[7] Bostan I, Dulgheru V and Ciobanu R. 2016 Elaboration and research of planetary precessional multiplier type K-H-V *IOP Conf. Ser.: Mater. Sci. Eng.* **147** 012002

[8] Lustenkov M E and Lustenkova E S 2020 Load Capacity of Spherical Roller Transmission with Double-Row Pinion *IOP Conf. Ser.: Mater. Sci. Eng.* **795** 012020

[9] Schenk H 1968 *Theories of Engineering Experimentation* (Moscow)

[10] Lustenkov M E and Lustenkova E S 2019 Spherical Roller Transmission with Double-Row Pinion: power calculations and determination of efficiency *Bulletin of Bryansk State Technical University* **5** 32–43

Simulation of surface work-hardening process during controlled inertial impact treatment of parts made of high strength cast iron

V M Pashkevich[1] and V V Afanevich[2]
[1]Belarusian-Russian University, Mira Avenue 43, Mogilev, 212000, Republic of Belarus
[2]Department of Mechanical Engineering Technology, Belarusian-Russian University, Mogilev, Mira Avenue 43, Mogilev, 212000, Republic of Belarus

E-mail: pvm@bk.ru; vitas-ik@tut.by

Abstract. The paper deals with the results of simulation of elastoplastic impact of the deforming ball with the surface during inertial impact treatment. The deforming ball is made of hardened steel ShKh15 (GOST 801-78), and special high strength spheroidal graphite cast iron is used as the treated material. The paper provides recommendations on conditions for inertial impact treatment of surfaces.

1. Introduction

At present, a number of different methods are used for finishing and hardening treatment of inner cylindrical surfaces of parts made of high strength cast iron. Apart from reducing roughness, they produce a work-hardened layer, which increases the surface wear resistance. Shot peening is a widely used surface treatment method. Although shot peening is a relatively simple process, it has a number of disadvantages, including poor controllability and the necessity of using the abrasive recovery system and a special chamber. Inertial impact treatment (IIT) can be used as an alternative to shot peening as it is devoid of the above disadvantages. The main advantage of IIT is the ability to provide controllability of the technological process, both by creating a partially regular microrelief of the surface and by obtaining the required hardening properties. The tool used for IIT can be mounted on the most common types of metalworking machinery – lathes, drilling and milling machines.

A schematic representation of the tool used for IIT of cylindrical holes is shown in figure 1 [1]. When the tool is in operation, the rotational motion from the workpiece is transmitted through collet 9 with friction rings 10 to the ball-type reducer comprising drive link 5 and driven link 4, the latter serves as a supporting element for deforming balls 3. The supporting element sets the deforming balls in motion by interacting with disk with radial grooves 2. The balls move in the radial direction perpendicular to the surface and strike it. Concurrently, a change in the frequency of rotation of the workpiece leads to a change in the deforming ball velocity in the direction of the surface being treated at the moment of impact. When the feed motion of the tool changes in the axial direction, the entire surface is covered with dents produced by deforming balls striking the surface. These multiple dents have regular or partially regular characteristics.

Content from this work may be used under the terms of the Creative Commons Attribution 3.0 licence. Any further distribution of this work must maintain attribution to the author(s) and the title of the work, journal citation and DOI.
Published under licence by IOP Publishing Ltd

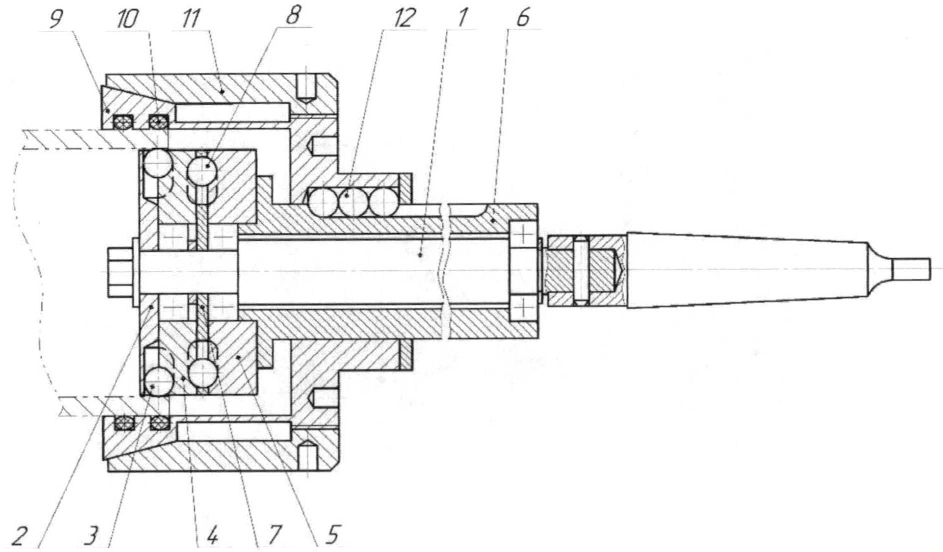

Figure 1. Inertial impact burnishing tool for holes.

To study the IIT process, a computer model simulating the impact of the deforming ball with the surface it strikes has to be created. The present paper is devoted to creating the simulation model and studying the IIT process characteristics.

The paper is aimed at building a computer model and performing numerical analysis of the process of inertial impact treatment of cylindrical holes in products made of high strength cast iron.

The methods used in this study are numerical methods (including the finite element analysis performed within SolidWorks environment) and methods of numerical approximation.

2. Results and discussion

By using the finite element analysis in SolidWorks environment, a model of elastoplastic impact of the deforming ball with the surface was developed.

The simulation of objects made of special high strength cast iron with spheroidal graphite, which has characteristics similar to VCh 60 cast iron (GOST 7293-85), was performed. The deforming balls were made of tempered bearing structural steel ShKh15 (GOST 801-78) with a diameter of 10 mm.

In the model, the velocity of the deforming ball at the moment of impact with the surface was varied.

During the simulation, the displacement of the material of the treated surface and arising residual stresses were studied.

It was assumed that the ball and the surface to be treated were perfectly smooth (no roughness), and shape and location errors did not occur. The impact was produced in the direction strictly perpendicular to the surface being treated; the surface at the point of contact was flat. The surface was subjected to a single impact; possible repeated impacts were not taken into consideration.

Enhanced ten-noded tetrahedral elements were used as finite elements (figure 2).

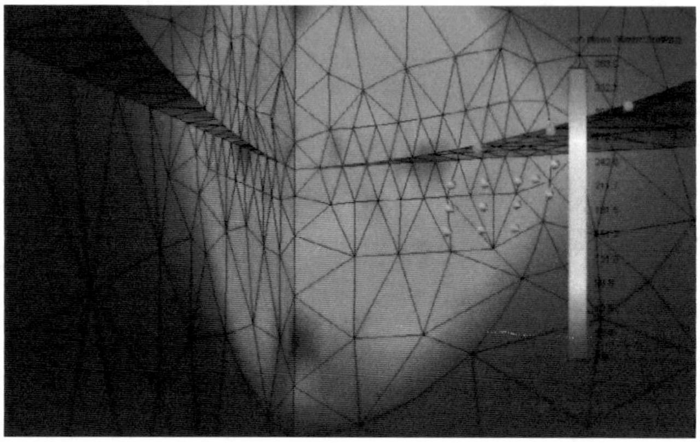

Figure 2. Results of finite element simulation of elastoplastic impact.

Before starting the calculations, the duration of interaction between the deforming ball and the surface was determined numerically (from the moment the ball strikes the surface to the moment the surface returns to its original state and the output parameters of the model are stabilized). For most numerical experiments, this time was assumed to be equal to 6×10^{-6} seconds.

The results obtained from the computer model simulation have shown that the deforming ball leaves a crater-like impression (dent) on the surface (figure 3, 100x magnification along the y-axis), which is in good agreement with the experimental data [2, 3].

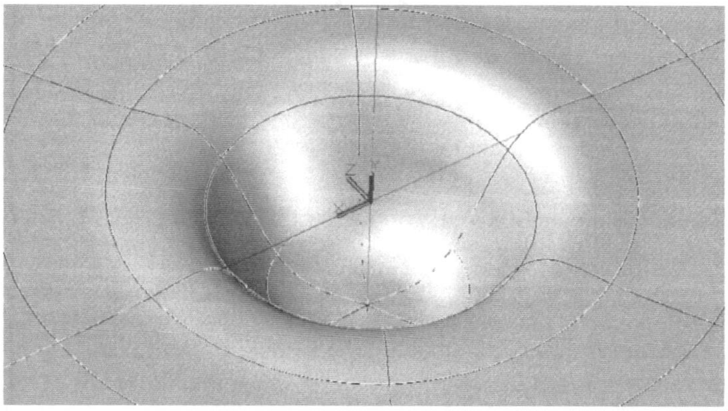

Figure 3. Deforming ball dent.

As can be seen in the figure 3, after the deforming ball strikes the surface, a number of phenomena related to formation of a new microrelief occur.

Firstly, the material in the central region of the ball dent (indentation) starts to flow; the surface tends to return to its original shape and becomes bulged to a greater or lesser extent. The amount of bulging is especially noticeable in ductile materials (steel) and less visible in cast iron (in the latter case, it is relatively larger for weak impacts).

Secondly, the material is squeezed out of the indentations into peripheral areas, which leads to formation of budging in the form of an annular ridge; the outer sides of the ridge asymptotically go down to the level of the initial surface.

Thirdly, under real conditions, when the impact produced by the ball in the radial direction occurs simultaneously with the rotation of the workpiece in the tangential direction, the dent caused by the impact becomes "blurred", the surface is partially smoothed out, and the shape of the indentation becomes elliptical, elongated.

These three phenomena lead to the formation of the surface microrelief as a combination of crater-like areas [4, 5].

After IIT, the value of residual stresses in the depth of the workpiece is characterized by nonlinear dependence (figure 4), and at a noticeable distance from the treated surface this value tends to zero. For this reason, it is rather difficult to estimate the depth (borders) of the hardened layer.

For practical purposes, the following estimation criterion can be used.

The depth of work-hardening is the depth at which residual stresses exceed their maximum allowable value for the as-received material. For stabilized products made of ductile high strength cast irons, residual stresses should not exceed 20-40% of their yield strength. For example, the yield strength σ_y for VCh 60 cast iron is 370 MPa. For such materials, a magnitude equal to 30% of the yield strength, i.e. 111 MPa, was taken as the maximum allowable value of residual stresses.

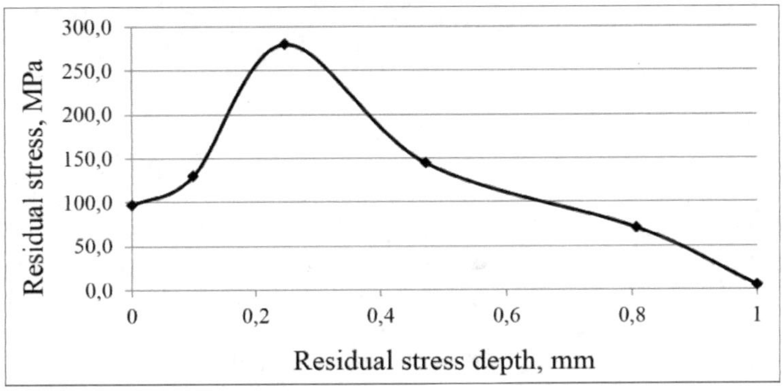

Figure 4. Distribution of residual stresses in the material under the point of impact (deforming ball material - ShKh15; workpiece material – high strength cast iron; ball diameter - 10 mm; ball velocity - 1.5 m/s).

The relationship between the depth of residual stress penetration into the material and the velocity of the deforming ball at the moment it strikes the surface, obtained from the simulation, is shown in figure 5.

The residual stresses of the work-hardened area are located in the subsurface layer both on the sides of the indentation and on the sides of the ridge of the ball dent.

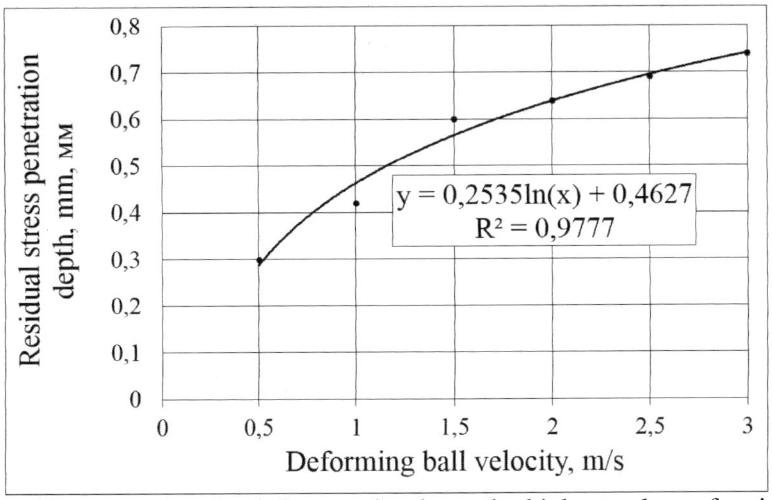

Figure 5. Relationship between residual stress penetration depth and deforming ball velocity.

Hence, as the simulation results show, the highest values of residual stresses are observed both in the central region of the indentation (microrelief below the initial surface) and in the area of the ridge itself (microrelief above the initial surface). Besides, residual stresses are noticeable in areas outside the dent, below the initial surface of the workpiece adjacent to the ridge, especially in case of high velocity impacts. This result is illustrated by an overlay graph showing the relationship between the location of residual stresses on the surface and the distance from the point of impact (figure 6).

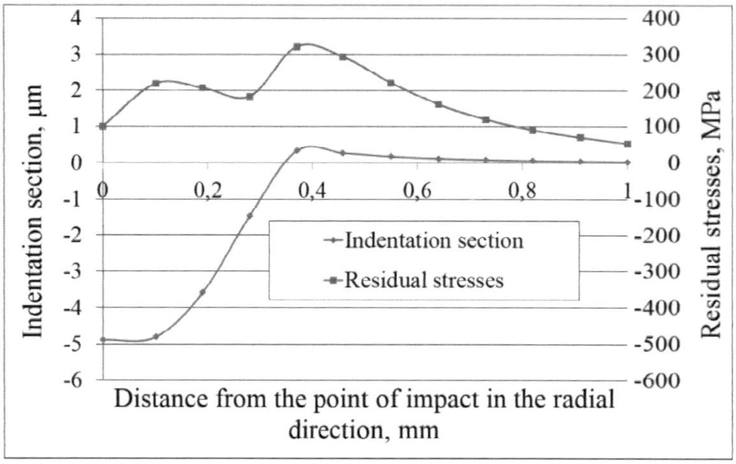

Figure 6. Relationship between location of residual stresses on the surface and distance from the point of impact superimposed on the half-section of the indentation.

As shown in figure 6, the residual stresses on the treated surface vary with distance from the point of impact, and the nature of these variations remains unchanged with different velocities of the deforming ball.

We believe that the diameter of the work-hardened area can be estimated if the formed surface microrelief is considered from the point of view of its future contact with the mating surface. Obviously, this contact will occur mainly in the area of the ridge, and at the beginning of the running-in process it will not affect the area of the indentation. The radius of this area corresponds to transition of the magnitude of residual stresses to values lower than the maximum allowable value for the as-received material (as indicated above, the magnitude of 111 MPa was taken as the maximum allowable value).

It is evident that residual stresses exceeding this value are located in a ring-shaped region. Figure 7 shows the relationship between the width of this region and the deforming ball velocity obtained from the simulation.

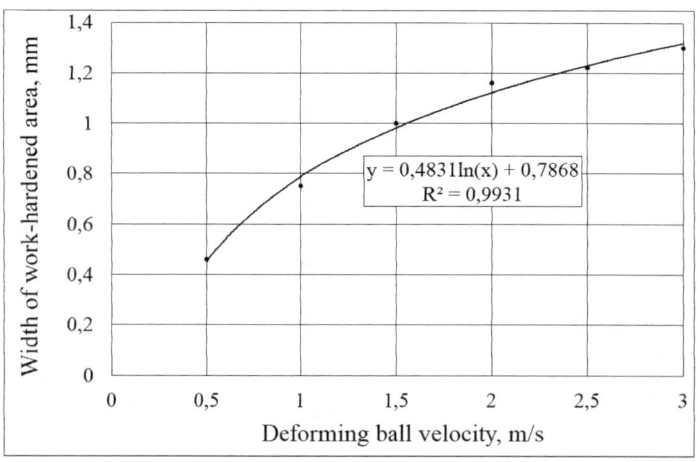

Figure 7. Relationship between width of work-hardened area with residual stresses and deforming ball velocity.

As can be seen from the graph, the relationship is nonlinear and tends to saturate. It is also obvious that a significant increase in the impact velocity does not lead to a noticeable increase in the diameter of the work-hardened area, which allows us to recommend the IIT conditions under which the deforming ball velocity of 1-3 m/s can be achieved. The results obtained are consistent with the experimental data. However, it should be noted that the above recommendations are valid only in terms of obtaining a specified value of residual stresses and the requirement to reduce surface roughness is not taken into account.

3. Conclusions

The results of the numerical simulation confirmed the hypothesis about the possibility of controlling the IIT-based process of high strength cast iron hardening.

The IIT process can be controlled by changing the frequency of rotation of the workpiece which causes a change in the deforming ball velocity. The density of the surface coverage with ball dents is regulated by changing the tool feed in the axial direction.

To perform IIT, it is recommended to provide conditions under which the deforming ball velocity at the moment of impact is in the range of 1-3 m/s and residual stresses with values of 250-300 MPa at a depth of 0.4-0.75 mm are generated. Residual stresses on the surface of the workpiece are concentrated in areas with a diameter of up to 0.8-1.3 mm.

4. References

[1] Pashkevich V M and Afanevich V V 2019 *Inertial impact burnishing tool* Patent BY No. 12381 4 (135) p 9

[2] Afanevich V V, Pashkevich V M and Popkovsky V A 2019 *Simulation of surface microrelief formation during inertial impact treatment J. Bel.-Rus. Univ. Bulletin* **4** 5–14

[3] Pashkevich V M, Afanevich V V and Maksimov K A 2019 Controlled inertial impact rolling of inner cylindrical surfaces Mechanical Engineering and Metalworking *Proc. International scientific and practical conference (Mogilev – Bobruisk, 10-11 September 2019)* ed M E Lustenkov (Mogilev: Belarusian-Russian University) p 48

[4] Kirk D 2013 Peening Impressions (Dents) *Coventry: Coventry University, The Shot Peener magazine* **27 (2)** 24-32

[5] Kirk D 2017 Work-Hardening During Peening *Coventry: Coventry University, The Shot Peener magazine* **31 (3)** 26-34

Effect of Glow Discharge Treatment on the Depth and Degree of Structural Phase Modification of Tool Steel Products

V M Shemenkov[1] and M A Rabyko[2]
[1]*Mechanical Engineering Technology* Department, Belarusian-Russian University, Mogilev, Belarus
[2]*Metal-Cutting Machines and Tools* Department, Belarusian-Russian University, Mogilev, Belarus

E-mail: Vshemenkov@yandex.ru; Belay-marina@ yandex.by

Abstract. The structure and phase composition of 1.1625, 1.2080 and 1.3343 tool steels (DIN) were studied and the research results are presented. The glow discharge treatment is shown to lead to dissolution and redistribution of the carbide phase, a change in the degree of crystal lattice distortion of the matrix phase, and a decrease in the residual austenite in the surface layer to a depth of 80 μm, which results in a 15% to 30% increase in the microhardness of the surface layers of materials.

1. Introduction

The surface layer is known to have a significant effect on the performance of products, and as a consequence, one of the modern world trends, especially in mechanical engineering, is to develop methods for increasing the wear resistance of surface layers of critical parts operating under conditions of tribomechanical and tribochemical effect.

The research was carried out as part of the *Physical Materials Science, New Materials and Technologies* State Program of Scientific Research of the Republic of Belarus, and its results show that treatment by glow discharge ignited in the atmosphere of residual atmospheric gases is one of the promising methods of increasing wear resistance of surfaces of the products made of metals and their alloys due to the structural phase modification of the surface layer [1 – 4].

2. Purpose and object of the research, and experiment description

The purpose of the research was to establish the effect of glow discharge treatment on the depth of structural phase modification of tool steels.

Samples of the following steels were selected as the object of research:

- 1.1625 steel (DIN), commercially melted. The samples were quenched in water at temperatures from 790° C (holding time: 1 h 30 min) followed by tempering at 250° C during 2 h;
- 1.2080 steel (DIN), commercially melted, quenched in oil at temperatures from 970° C (holding time: 1 h 30 min) followed by tempering at 180° C during 1.5 h;
- 1.3343 steel (DIN), commercially melted, quenched in oil at temperatures from1,220° C (holding time: 1 h 30 min) followed by threefold tempering at 560° C during 2 h.

To study the dynamics of the process, the samples were subjected to glow discharge treatment during 30 min using three modes: mode No. 1 - combustion voltage 1,000 V, current density 0.125 mA/m^2;

Content from this work may be used under the terms of the Creative Commons Attribution 3.0 licence. Any further distribution of this work must maintain attribution to the author(s) and the title of the work, journal citation and DOI.
Published under licence by IOP Publishing Ltd

mode No. 2 - combustion voltage 2,000 V, current density 0.250 mA/m^2; mode No. 3 - combustion voltage 3,000 V, current density 0.375 mA/m^2.

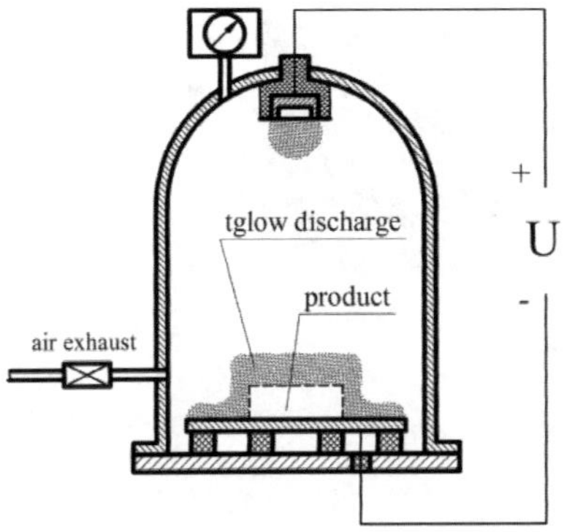

Figure 1. Schematic of the installation for modifying treatment.

3. Research methodology

In the research, an electron microscopic analysis of the surface layer of samples was conducted using a Tescan VEGA 2SBA scanning electron microscope equipped with an unexposed secondary electron detector, as well as an X-ray structural analysis of the metal phase composition using an automated X-ray system on the basis of a DRON-3M diffractometer with the application of monochromatic $Co_{k\alpha}$ - radiation. A durometric analysis of the surface layer was carried out using a Zwick Roell ZHV 1M hardness tester.

4. Research results and discussion

The electron microscopic analysis revealed that the structure of 1.1625 steel before glow discharge treatment is a classic martensitic structure with carbide inclusions of alloying elements up to 5 μm in size. The glow discharge treatment results in dissolution and redistribution of the carbide phase in the surface layer 60 to 65 microns deep (Figure 2)

Based on the X-ray diffraction analysis, it was found that in its initial state, 1.1625 steel contains martensite, retained austenite γ-Fe, as well as inclusions in the form of iron carbides (Fe_3C) (Figure 3). The crystal lattice parameter is close to the crystal lattice parameter of ferrite, which indicates an insignificant concentration of defects in the crystal lattice which are typical for martensite in steels (Table 1).

The glow-discharge treatment of 1.1625 steel with different combustion parameters does not result in a qualitative change in their phase composition (Figure 3). At the same time, after the treatment in mode No. 1, a decrease in the lattice parameter and an increase in the physical broadening of diffraction lines of the α-phase are recorded, which is due to an increase in the density of dislocations in the surface layer as a result of the treatment.

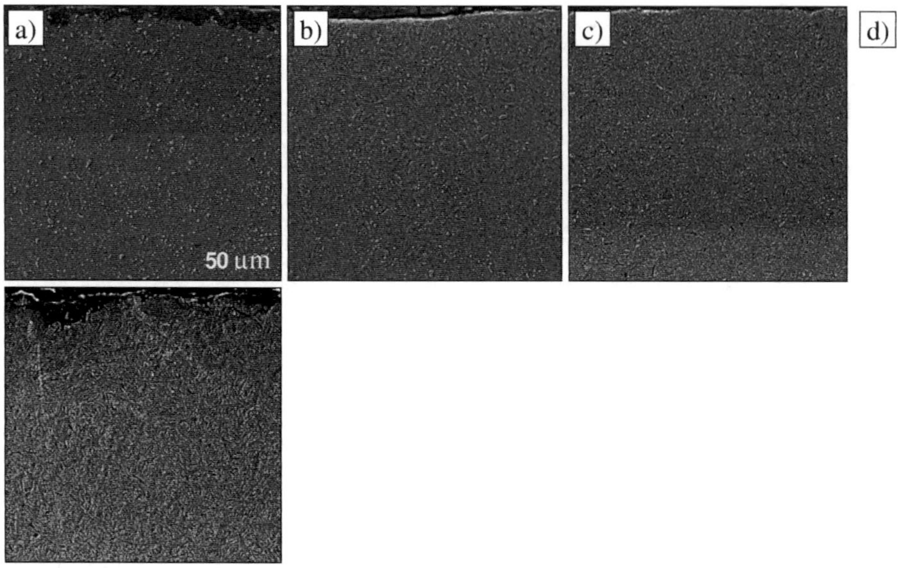

Figure 2. The structure of U8 steel: a – before processing; b – after processing mode 1; c – after processing mode 2; d – after processing mode 3.

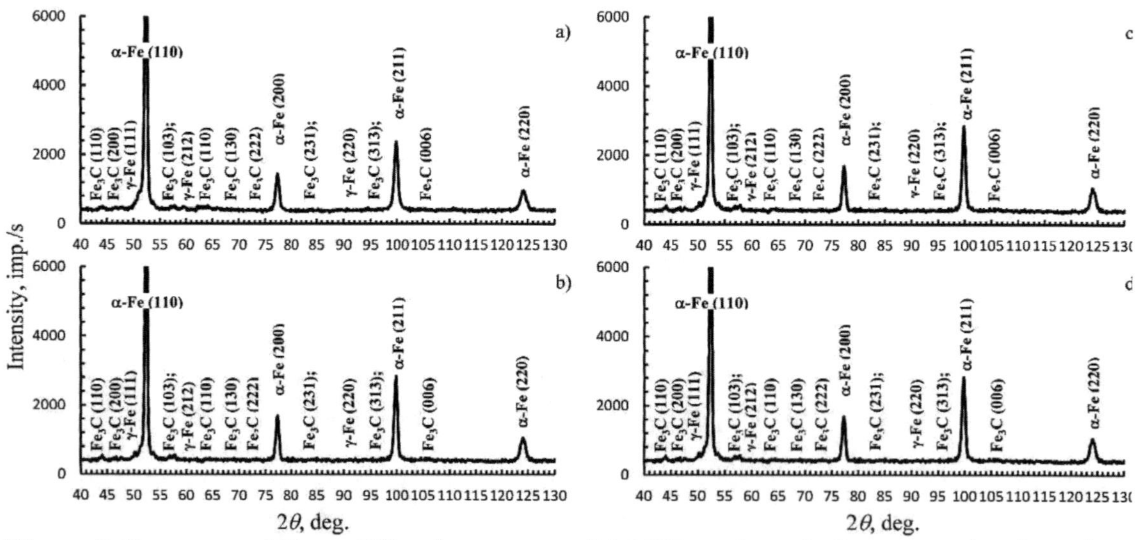

Figure 3. Fragments of X-ray diffraction patterns of 1.1625 steel: a – before processing; b – after processing mode 1; c – after processing mode 2; d – after processing mode 3.

Table 1. Phase composition, amount of retained austenite (A), crystal lattice parameter a, physical broadening β of diffraction lines and dislocation density ρ.

Sample	Phase composition	Amount of A, %	$a_{\alpha\text{-Fe}}$, nm		β_{110}, 10^{-3} rad	β_{220}, 10^{-3} rad	ρ, cm^{-2}
			experiment	standard [5]			
initial state	α-Fe, Fe3C, γ-Fe	5.7	0.28622	0.28664	5.24	22.14	$2.8 \cdot 10^{11}$

mode No.1	α-Fe, Fe3C, γ-Fe	4.1	0.28619	7.04	24.98	$5.1\cdot10^{1}$ 1
mode No.2	α-Fe, Fe3C, γ-Fe	4.8	0.28623	5.26	19.51	$2.8\cdot10^{1}$ 1
mode No.3	α-Fe, Fe3C, γ-Fe	4.4	0.28624	5.24	21.96	$2.8\cdot10^{1}$ 1

Prior to the glow discharge treatment, the surface layers of both 1.2080 steel and 1.1625 steel have a classical hardened martensite structure with carbide inclusions in the form of plates. The glow discharge treatment causes dissolution and redistribution of the carbide phase in the surface layer 80 to 85 μm deep (Figure 4).

The analysis of fragments of 1.2080 steel diffraction patterns (Figure 5) revealed that the glow discharge treatment of 1.2080 steel does not result in a qualitative change in their phase composition (Figure 4). At the same time, an increase in the lattice parameter of the α-phase is recorded after the glow discharge treatment, which is due to an increase in the concentration of interstitial atoms in the martensitic phase. The processing also leads to a decrease in the values of physical broadening of the α-phase diffraction lines, which is due to a decrease in the density of defects during the processing (Table 2).

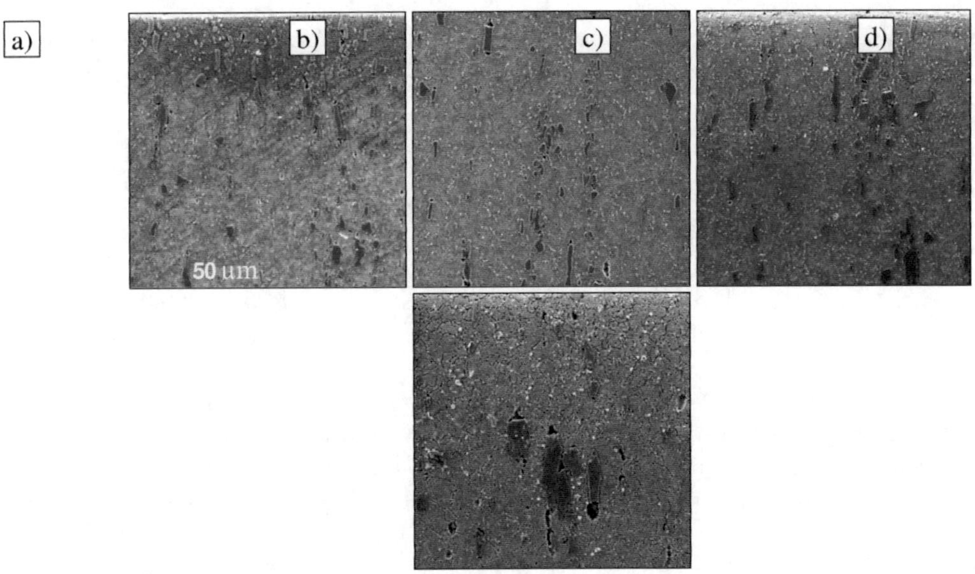

Figure 4. The structure of 1.2080 steel: a – before processing; b – after processing mode 1; c – after processing mode 2; d – after processing mode 3.

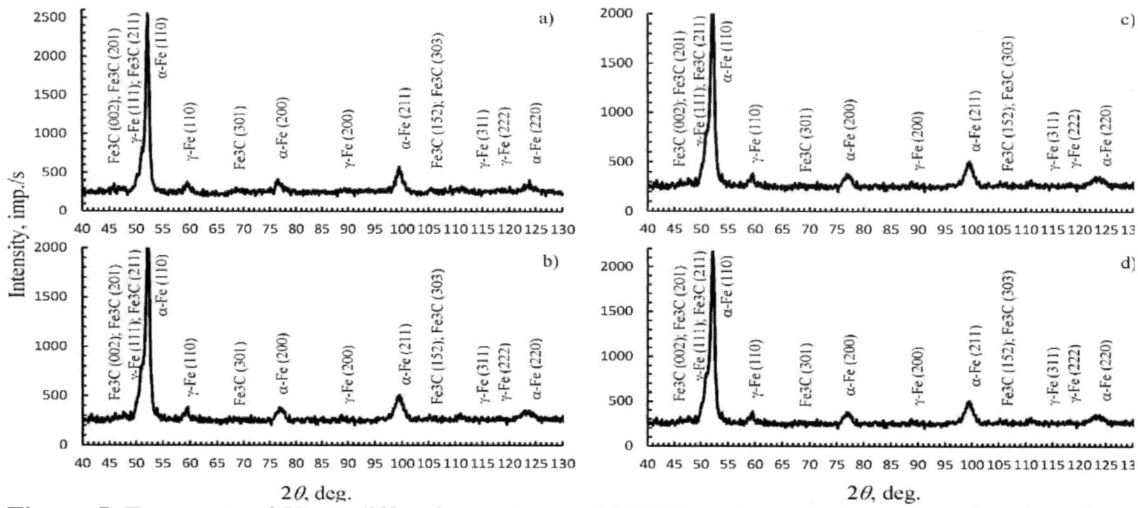

Figure 5. Fragments of X-ray diffraction patterns of 1.2080 steel: a – before processing; b – after processing mode 1; c – after processing mode 2; d – after processing mode 3.

Table 2. Phase composition, amount of retained austenite (A), crystal lattice parameter a, physical broadening β of diffraction lines and dislocation density ρ.

Sample	Phase composition	Amount of A, %	$a_{\alpha\text{-Fe}}$, nm		β_{110}, 10^{-3} rad	β_{220}, 10^{-3} rad	ρ, cm^{-2}
			experiment	standard [5]			
initial state	α, γ, M_3C	33.7	0.28734	0.28664	14.1	72.8	$2.1\cdot10^{12}$
mode No. 1	α, γ, M_3C	34.1	0.28742		18.0	100.6	$3.4\cdot10^{12}$
mode No. 2	α, γ, M_3C	31.1	0.28737		16.5	60.0	$2.9\cdot10^{12}$
mode No. 3	α, γ, M_3C	31.4	0.28745		18.0	71.8	$3.5\cdot10^{12}$

The structure of high-speed 1.3343 steel before it was subjected to glow discharge treatment, as well as the structure of the previous steels, corresponds to quenched martensite with finely dispersed carbide inclusions. The glow discharge treatment leads to dissolution and redistribution of the carbide phase in the surface layer 70 to 75 μm deep. This effect is most pronounced when processing in mode No. 3 (Figure 6).

The X-ray structural analysis of the surface of samples made of 1.3343 steel (Figure 7) showed that processing does not cause a change in the phase composition of steel. All steel samples under study include α-Fe, M_6C carbide and VC carbide (Table 3).

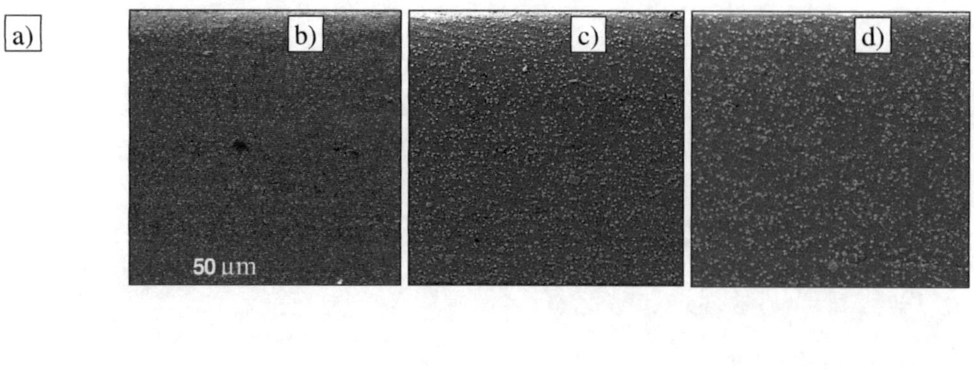

Figure 6. The structure of 1.3343 steel: a – before processing; b – after processing mode 1; c – after processing mode 2; d – after processing mode 3.

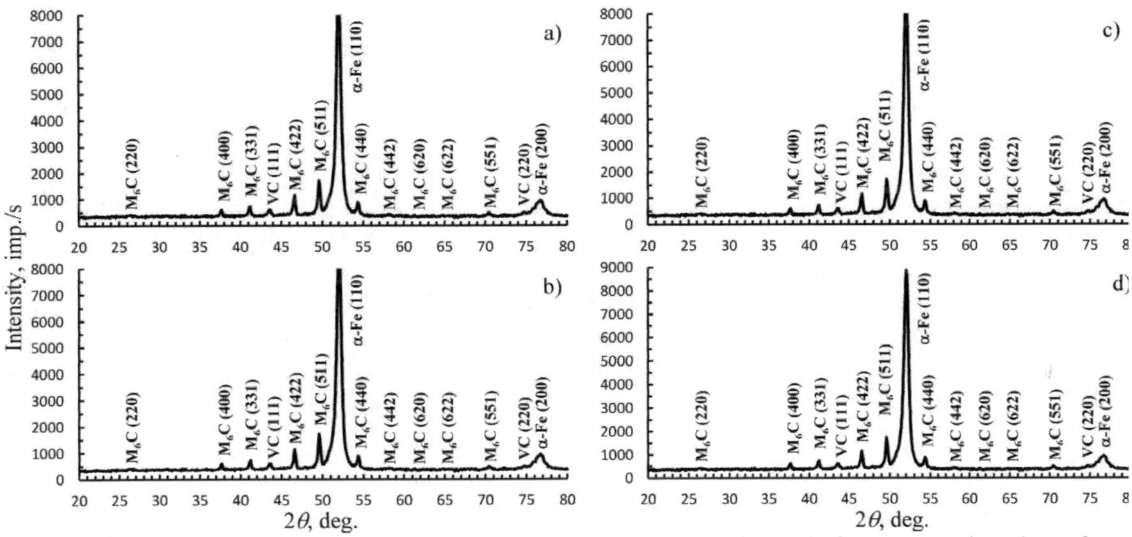

Figure 7. Fragments of X-ray diffraction patterns of 1.3343 steel: a – before processing; b – after processing mode 1; c – after processing mode 2; d – after processing mode 3.

The recorded broadening β of the diffraction lines is proportional to tgθ, which indicates that the broadening observed is mainly due to a high density of randomly distributed dislocations. Lower values of the physical broadening of the α-Fe diffraction line (220) recorded for the samples subjected to glow discharge in processing modes 2 and 3, indicate a decrease in the dislocation density in the surface layer of the 1.3343 steel samples.

Table 3. Lattice parameter a and physical broadening β of α-Fe diffraction lines.

Sample	a, nm		β_{110}, 10^{-3} rad	β_{220}, 10^{-3} rad
	experiment	standard [5]		
initial state	0.2882		15.5	75.1
mode No. 1	0.2880	0.28664	15.1	76.4
mode No. 2	0.2880		15.7	64.9
mode No. 3	0.2882		15.1	57.5

At the same time, the processing of samples in mode 3 leads to a significant increase in the physical broadening of the diffraction line (511) of M_6C carbide (β_{511} = 10.3 10^{-3}) as compared to unexposed samples (β_{511} = 6.4 10^{-3}), which is the evidence of the dissolution of carbide particles during processing.

As a result of the glow discharge treatment, a 15 – 30% change in the microhardness of the surface layers of materials is observed, in addition to structural changes. The results of the durometric analysis are shown in Figure 8.

The results shown in Fig. 2–7 are evidence that the data of the electron microscopic and durometric analyses correlate sufficiently well with each other and allow for a conclusion that the depth of the modified layer of tool steel products subjected to glow discharge processing depends significantly on the presence of alloying elements in them affecting the electrophysical properties, namely, the specific resistance.

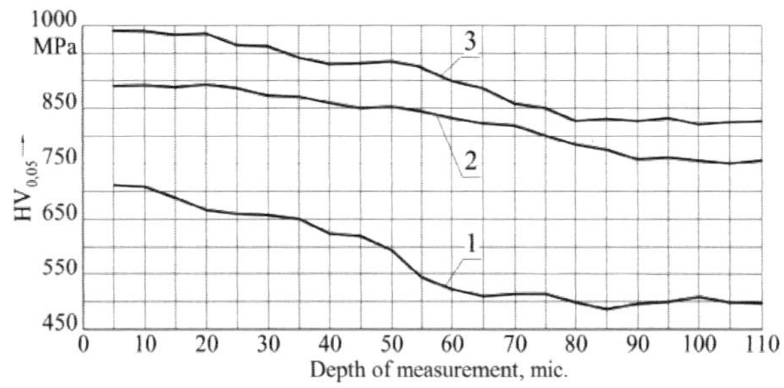

Figure 8. Microhardness of the surface layer through the depth after glow discharge treatment: 1 – 1.1625 steel; 2 – 1.2080 steel; 3 – 1.3343 steel.

5. Conclusions

The glow discharge treatment of tool steels leads to the dissolution and redistribution of the carbide phase, the change in the degree of crystal lattice distortion of the matrix phase, and the decrease in retained austenite due to the polymorphic transformation γ–Fe $\rightarrow$ α–Fe in the surface layer to a depth of 80 μm. These changes, in their turn, ensure a 15 – 30% increase in the microhardness of the surface layers of materials.

References:

[1] Shemenkov V M, Trukhachev F M, Obidina O V, Belaya M A, Shemenkova A L 2018 *Technique of strengthening products made of metal or alloy, or superhard or graphite-containing material* Patent BY No. 22048 C1 **3**.

[2] Grigoriev S N 2009 *Methods for increasing the durability of cutting tools: a textbook for students of technical colleges* (Moscow: Mashinostroenie) p 368.

[3] Shemenkov V M [et al.] 2017 *Structural Phase Modification of Tool Materials by Using a Glow Discharge* (Mogilev: Bel.-Rus. Univ.) p 270.

[4]. Grigoriev S N, Tabakov V P, Volosova M A 2011 *Technological methods of increasing the wear-resistance of contact pads of the cutting tool* (Stary Oskol: TNT) p 372

[5] Gorelik S S, Skakov Y A and Rastorguev L N 2002 *X-ray and Electrooptical Analysis (*Moscow: MISIS)

Surface treatment by combined processes of cutting and magnetodynamic rolling

D M Svirepa and A S Kruchynina
Belarusian-Russian University, Mira Avenue 43, Mogilev, 212000, Republic of Belarus

E-mail: svirepa.dmitry@gmail.com

Annotation. Technologies for mechanical treatment of the inner cylindrical surface of parts using combination tools have a number of advantages over traditional methods. The combination of two technological processes in one operation (cutting and magnetodynamic rolling) leads to increased production efficiency, reduced cost of parts manufacturing and improved quality characteristics of the machined surface. The developed designs of tools for combined treatment with various magnetic modules (with radial and axial arrangement of the magnetic system) have a number of specific features and advantages, which makes it possible to expand the field of application of the combination tools.

1. Introduction

Methods of surface plastic deformation have been widely used in industry [1-4]. Technologies of surface treatment by combining the cutting operation and the method of surface layer modification by magnetodynamic rolling are used to obtain high performance surfaces and reduce the time required to manufacture parts, which leads to increased productivity and improved competitiveness of products.

The technology of finishing and strengthening treatment of holes using magnetodynamic modules has a number of advantages over classical methods of treatment of internal cylindrical surfaces, e.g. it helps obtain high-level quality characteristics of the surface layer of treated parts by improving durability as well as wear and corrosion resistance of the product in the shortest amount of time [5].

2. Results and discussion

The technology of surface treatment by combined processes of cutting and magnetodynamic rolling is used to perform finishing and strengthening treatment of holes in cylindrical parts such as cylinders, bushings, rings with a diameter of over 60 mm. The treatment is performed using a tool with boring and magnetic-dynamic modular components, which are mounted coaxially on a single mandrel. To carry out this technological operation (the combined treatment), the combination tool is preliminarily adjusted for the proper depth of cut taking into account the amount of allowance for finishing or semi-finishing (0.1 – 0.5 mm). For this purpose, a jig bush can be used. Cylindrical part 8 is fixed in a device preliminarily mounted on the machine table (drilling and milling machines) or in a chuck (lathes). Figure 1 shows a schematic diagram of the combined treatment.

Content from this work may be used under the terms of the Creative Commons Attribution 3.0 licence. Any further distribution of this work must maintain attribution to the author(s) and the title of the work, journal citation and DOI.
Published under licence by IOP Publishing Ltd

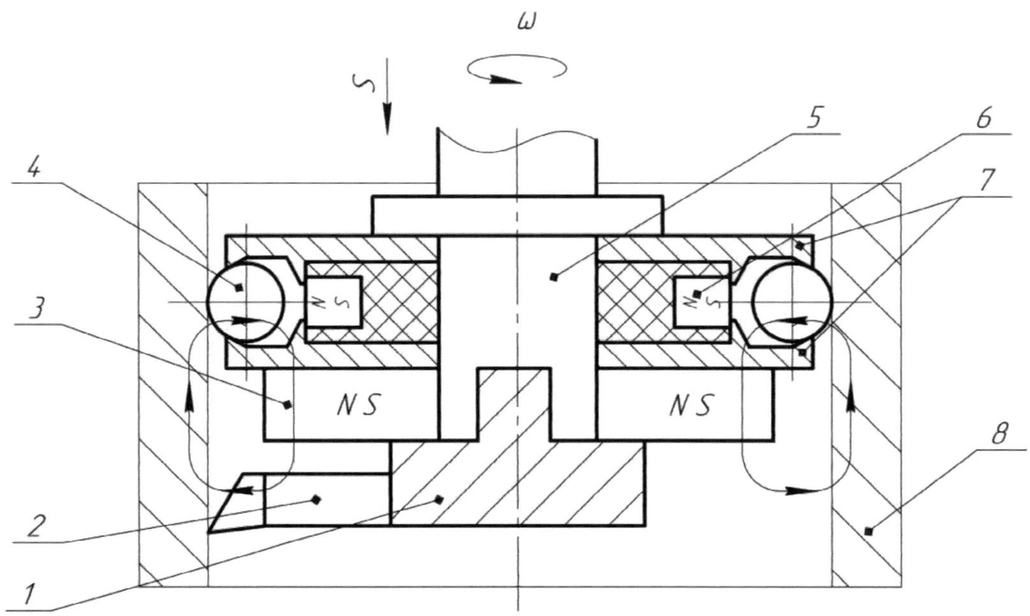

Figure 1. Combined cutting and magnetodynamic rolling treatment technology.

The rotational motion ω in the range from 350 to 500 m/min is imparted to mandrel 5 fixed in the spindle (drilling and milling machines). In lathes, rotation is imparted to workpiece 8. The feed motion S (0.02-0.16 mm/rev; 22.4-256 mm/min) is transferred either to the part or to the tool depending on the type of machine used for treatment.

Cutting module 1 includes cutting tool 2 for carrying out the cutting operation and ensuring the dimensional accuracy of the workpiece. Adjustment to the required diametrical dimension is performed using a boring head. Besides, cutting module 1 has a source of additional magnetic field 3 that carries out two tasks: it removes the chips formed during the boring operation from the zone deformed by the magnetic module and performs pre-magnetization of the deformation zone.

The magnetodeforming module is designed to perform finishing and strengthening treatment and contains a magnetic system with radially oriented magnetic field sources in the form of permanent cylindrical magnets 6 and a number of deforming balls 4. The deforming balls are freely located in the annular chamber (cavity) formed by two disks 7 and are set in motion by the joint action of centrifugal force and concentrated magnetic flux (energy). The balls perform oscillatory movements striking the surface of the workpiece at regular intervals [6, 7].

In order to intensify the process of the surface layer modification and increase the durability of the boring component of the tool, an additional speed of rotation in the direction coinciding with the rotation of the tool is imparted to the workpiece, which reduces the relative speed of the tool with respect to the workpiece. In this case, the boring speed is equal to the difference between the speed of rotation of the tool and that of the workpiece and can be reduced to 200-250 m/min, and the rotational speed of the rolling module can be increased to 500-1000 m/min. Composite materials modified by cubic boron nitride can be used as tool materials [8].

3. Materials, equipment and methods

The main requirement in critical machine parts manufacturing is to achieve the accuracy specified on the drawing. To determine the accuracy of the surface treatment technique by combined processes of cutting and magnetodynamic rolling, a statistical and computational method and advantages of statistical and probabilistic as well as computational and analytical methods of machining accuracy calculation were used.

A batch of 80 workpieces was used to conduct the experimental studies.

The parameters of the workpieces used in the experiments were as follows:

- dimensions ($D \times d \times l$), mm – 120 × 99.6 × 50 mm;
- workpiece material: steel 20 (180–200 HB);
- hole surface roughness *Ra* 10.4–11.7 µm.

For subsequent treatment, the batch of workpieces was divided into two parts, forty pieces each.

The treatment of both batches of workpieces was performed on a HAAS ST-20Y machine.

The boring operation of the first batch of workpieces was carried out by means of a tool with a cutting module; the magnetodynamic module of the tool had been removed.

The second batch of workpieces was machined with a tool designed for combined treatment by cutting and magnetodynamic rolling (with both cutting and magnetodynamic modules).

Both batches were machined under the same conditions:

- cutting speed 703 m/min (frequency of rotation 2240 min^{-1});
- axial feed of the tool S = 20 mm/min;
- depth of cut $t_c = 0.2$ mm;
- the number of tool passes: 1;
- metalworking fluid: industrial oil 45.

The diameter of the machined parts was measured using a dial bore gauge with a graduation of 0.001 mm GOST 9244-75.

First, the tool was adjusted to achieve the desired geometry of the workpieces (tolerance range). The batches of workpieces were sequentially machined. After that, they were grouped according to the diametrical dimensions obtained, and the average value X_{ia} of each corresponding interval and the frequency of repeated dimensions were calculated.

Based on the research results, a distribution plot of diametrical dimensions *X* versus the frequency of repeats ω was created (figure 2).

To test the hypothesis of normal distribution, the Pearson Chi-Square and Kolmogorov goodness-of-fit tests were performed and the following values were obtained χ^2_o=1.27; $\lambda_o = 0.317$. The goodness of fit criteria for the second batch of parts were as follows: χ^2_o= 0.7; $\lambda_o = 0.17$. Since the observed criteria χ^2_o and λ_o are less than the critical values, there is no reason to reject the hypothesis of a normal distribution [9].

The experimental studies have demonstrated that the diametrical dimensions of the workpieces obtained after machining follow the Gaussian distribution. After the parts are machined with the combination tool, they are still within the tolerance range.

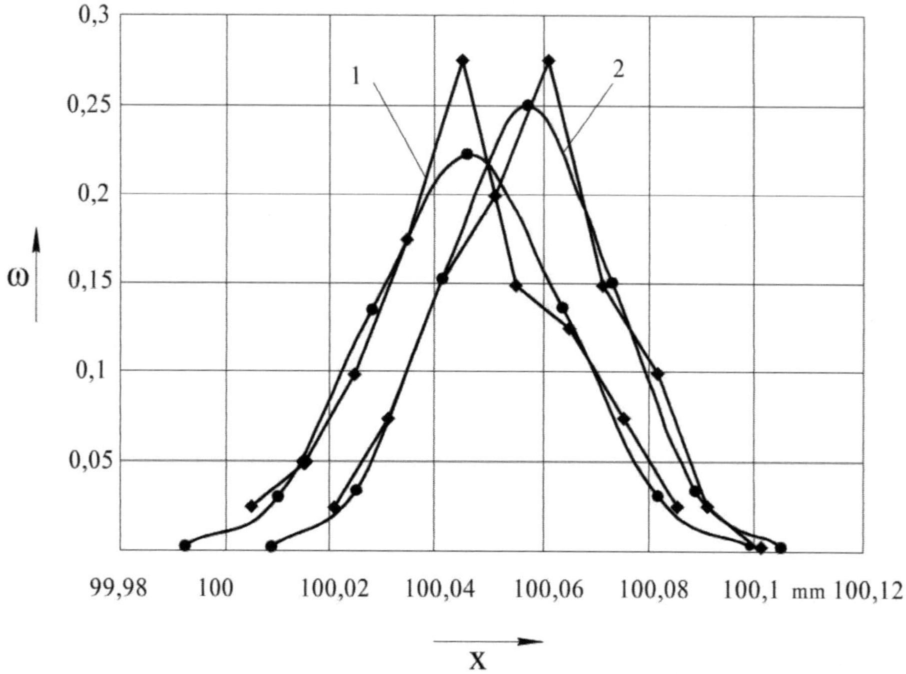

1 – diametrical dimension distribution plots after boring; 2 – diametrical dimension distribution plots after combined cutting and magnetodynamic rolling treatment

• theoretical curves; ♦ empirical curves

Figure 2. Study of accuracy after boring and after combined cutting and magnetodynamic rolling treatment.

After performing experimental studies, graphs of the dependence of the change in the original size of the part on the rotational speed of the tool was created after the operations of boring and combined cutting and magnetodynamic rolling performed with a tool with radially oriented magnetic field sources (figure 3).

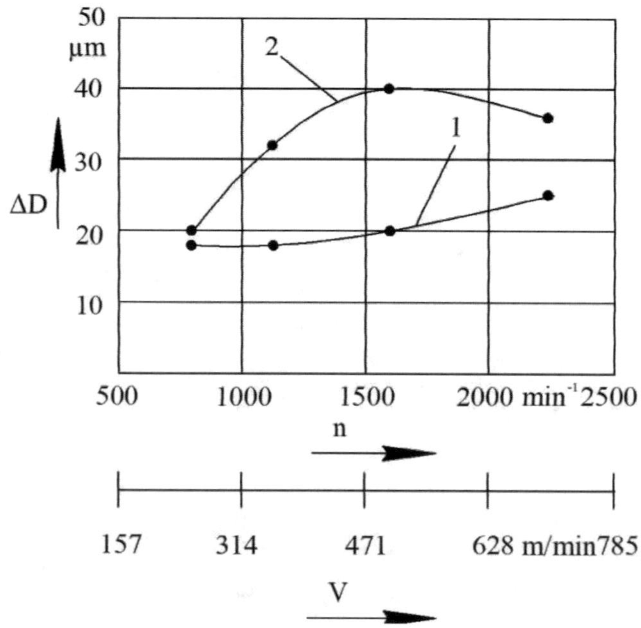

1 – boring of steel 20 (180-200 HB) performed by means of cutting module;
2 – treatment performed by means of combination tool designed for cutting and magnetodynamic rolling with radially oriented magnetic system of steel 20 (180-200 HB)

Figure 3. Dependence of change in original dimension of part on rotational speed of tool.

4. Conclusions

Summing up the results, it can be concluded that the change in the original dimension of the part occurring as a result of mechanical treatment by combined processes of cutting and magnetodynamic rolling is equal to the sum of two values, the amount of the allowance to be removed by the cutting module and the amount of the allowance to be modified by magnetodynamic rolling, and lies within the range 18-20 μm, which can be used to provide recommendations on calculation of the allowance for the combined treatment.

The developed technology of surface treatment by combined processes of cutting and magnetic-dynamic rolling provides the following advantages:

– surface roughness *Ra* decreases significantly (from 10.4-10.7 μm to 0.2-1.7 μm);
– the surface layer is strengthened to a depth of 0.05-1 mm;
– the original diametrical dimension of the part is changed during combined cutting and magnetodynamic rolling treatment;
– treatment of parts with hardness up to 50 ... 55 HRC can be performed;
– machining efficiency is increased by a factor of 2-5.

References

[1] Nagorkin M N, Fedorov V P and Kovalyova E V 2019 Process control of forming the surface quality parameters of parts according to a desired law by methods of surface plastic deformation *Int. Conf. on Aviamechanical Engineering and Transport AviaENT 2019* **188** 258–263
[2] Dovgalev A M, Svirepa D M and Ryzhankov D M 2008 Classification of tools for magnetodynamic strengthening *Belarusian-Russian University Bulletin* **2** 30–38
[3] Pilipenko O V, Golenkov V A, Radchenko S Y and Dorohov D O 2018 Numerical athematical Simulation of Penetration of Indenters of Different Shapes During Strengthening Machine Parts via Local Loading of Deformation Zone *Int. Symp. on Engineering and Earth Sciences ISEES 2018* **177** pp 568–573
[4] Gryadunov I M, Golenkov V A, Pilipenko O V and Radchenko S J 2017 Hardening treatment by plastic deformation under conditions of the integrated local loading of a deformation zone *Int. Journal of Applied Engineering Research 2017* **12** pp. 11094-11100
[5] Dovgalev A M and Svirepa D M 2014 Technology of magnetodynamic rolling and its implementation in mechanical engineering *Proc. Materials, Equipment and Resource-Saving Technologies* (*Mogilev, Belarus, 24–25 April 2014*) (Mogilev: BRU) pp 10–15
[6] Svirepa D M and Semionova A S 2019 Optimization of the number of magnetic field sources in tools used for combined processes of cutting and magnetodynamic rolling *Proc. New materials, equipment and technologies in industry* (*Mogilev, Belarus, 24-25 October 2019*) (Mogilev: BRU) p 40
[7] Svirepa D M and Semionova A S 2019 Influence of methods of combined treatment by cutting and magnetodynamic rolling on surface roughness *Belarusian-Russian University Bulletin* **2** 34–43
[8] Kudryashov E A, Smirnov I V and Khizhnyak N A 2018 Selection and evaluation of the efficiency of cutting inserts made of various cutting tool materials *Int. Conf. on Aviamechanical Engineering and Transport AviaENT 2018* **158** 226–229
[9] Vadzinskij R N 2001 *Probability Distributions Handbook* (St.Petersburg: Science)

IOP Conf. Series: Materials Science and Engineering 1118 (2021) 012011 doi:10.1088/1757-899X/1118/1/012011

Effect of hot isostatic pressing on the quality of metal parts obtained by selective laser melting

A V Zhdanov*, **A A Voznesenskaya and A S Raznoschikov**

Vladimir State University, Mechanical Engineering Technology Department, 87 Gorky Street, Russia

*E-mail: zhdanov@vlsu.ru

Abstract. In this work, an approach to improving the quality of material obtained by the method of selective laser melting is considered. The analysis of powder material of stainless steel is carried out. A number of samples were obtained with subsequent HIP-processing. The structure was analyzed, and the hardness of the obtained samples was measured before and after HIP treatment.

1. Introduction

One of the topical areas of application of additive technologies is medicine and the production of unique medical products. This area has a need for individual and structurally complex parts that work in extremely critical places and areas. In addition to strength, an important parameter is the density of the resulting structure of the part [1,2]. Selective laser melting (SLM) is one of the additive manufacturing technologies, a feature of which is the piece production of complex parts. SLM assumes layer-by-layer growing of a powder metal part in an inert gas environment. In the process, the part is heated in the area of laser action and non-equilibrium cooling. This leads to uneven shrinkage of the metal. In the process of manufacturing a part by the SLM method, internal residual stresses associated with the presence of a temperature gradient, the quality of the powder material, as well as incorrectly selected parameters for growing parts are inevitable. This leads to cracks and pores. [3]. The research results [4, 5] show the influence of certain parameters of the part growing on the final roughness and porosity of the product surface. These defects are unacceptable in the manufacture of critical parts for medical purposes in contact with biological tissue. Their elimination is possible not only by optimizing the SLM parameters, but also by post-processing [6]. Hot isostatic pressing (HIP) is a successful method for reducing microporosity in detail. HIP is a well-known process for reducing porosity and producing relatively large, bulky and dense parts. The process consists in supplying an inert gas under high pressure when heated to high temperatures in a special working chamber. The pressure is isostatic and therefore preserves the geometry of the component. Simultaneously, high temperature and pressure eliminate internal pores or defects [7, 8]. Reducing the size and number of pores in the final product is an urgent task.

2. Materials and methods

The production of samples took place on an industrial machine Concept Laser M2 cusing. This unit operates on the technology of selective laser melting and is designed for the production of parts of complex geometric shapes from reactive and non-reactive metal powders. Stainless steel powder with the following chemical composition was used as a starting material:

Content from this work may be used under the terms of the Creative Commons Attribution 3.0 licence. Any further distribution of this work must maintain attribution to the author(s) and the title of the work, journal citation and DOI.
Published under licence by IOP Publishing Ltd

Table 1. Chemical composition of the powder material.

Mass fraction of elements, %									
Fe	C	Si	Mn	Cr	Ni	Mo	S	P	O
base	0.07	0.75	2	16-18	10-14	2-3	0.02	0.03	0.10

Investigation of the particle size distribution was carried out by methods of optical and scanning electron (SEM) microscopy.

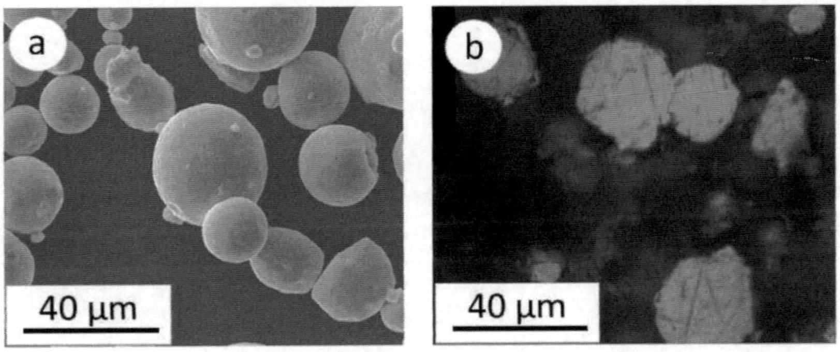

Figure 1. Image of powder particles: a - SEM, b - optical microscopy.

In figure 1 (b) it can be seen that the particles inside have a continuous structure and the pores are completely absent. Most of the particles are spherical, deformed particles are rare, the average fraction size is 40 µm. The powder material was exposed to laser action. The laser radiation wavelength was 1.06 µm. The scanning speed was varied from 500 to 1600 mm / s. Layer-by-layer melting occurs according to the following strategy: the grown layer of the part is fragmented into square areas, called "islands" with a size of 5 x 5 mm [9]. Laser melting of each individual "island" is performed. The fragments are melted according to a predetermined algorithm that allows localizing the internal stresses of the metal in a small area. Each "island" is processed by hatching, perpendicular to the neighboring areas, which allows stresses to be reduced to a minimum. When the next layer is fused, the dividing step shifts, the direction of the hatching inside the "islands" is rotated, providing a dynamically changing hatch pattern. All "islands" are fused with each other by completely melting the edges of the "island" - this allows to achieve the most homogeneous and dense structure with minimum porosity [10, 11]. The production of prototypes was carried out in a nitrogen atmosphere.

A group of samples was selected for further processing by hot isostatic pressing. HIP-processing was carried out using an isostatic press, which has the ability to process samples with stepwise heating up to 2000 C at a pressure of up to 2000 Bar. The HIP treatment process took place under the following operating parameters: operating temperature - 1000 ° C, operating pressure 1500 bar, temperature rise rate and fall rate - 6 seconds, holding time - 7 hours.

3. Results

The structural-phase composition was studied on a D8 Advance (Bruker). The material, which was not subjected to isostatic pressing, corresponded to a solution of chromium, nickel, and other components in iron, without the precipitation of other individual phases (Figure 2).

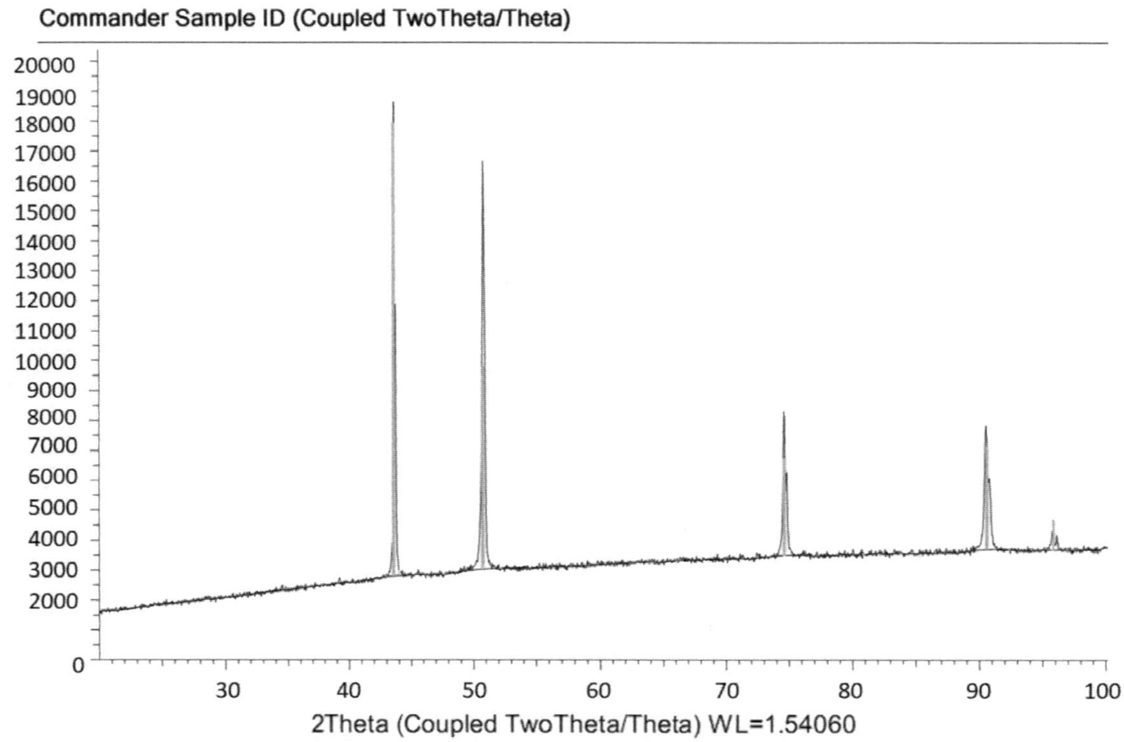

Figure 2. Structural phase analysis of powder samples.

The study defects are revealed by obtaining a thin section and etching of the transverse and longitudinal surface of the sample, followed by examination by means of optical microscopy. The initial samples have cellular structure contours typical for the SLM technology [12]. After processing, the samples have a mosaic structure typical of HIP [13]. Image acquisition and analysis of the porous structure was performed using an Altami microscope and AltamiStudio 5.0 software.

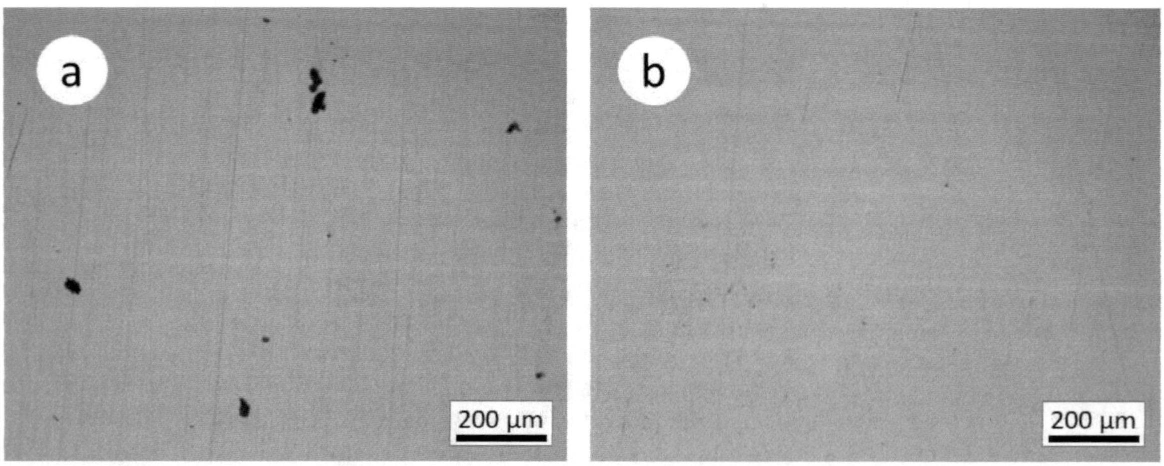

Figure 3. Microstructure of the obtained samples: a - the original sample, b - after HIP.

It was found that pores up to 8 µm in size were observed in the initial samples. The use of HIP reduces the maximum pore size to 1 micron, while the porosity of less than 1 micron is completely tightened and the material has a more uniform structure.

Hardness was measured at random points in planes 1, 2, and 3 (Fig. 4). Before HIP in area 1, the average hardness is greater than the average hardness in planes 2 and 3, this is due to the inherent anisotropy of mechanical properties in SLM, caused by the build strategy. In Figure 4, the growth trajectory Z is the direction of heat removal to the build substrate. Accordingly, directional crystallization of the melt pool is observed, which leads to an increase in hardness in the direction of solidification, rather than in planes 2 and 3.

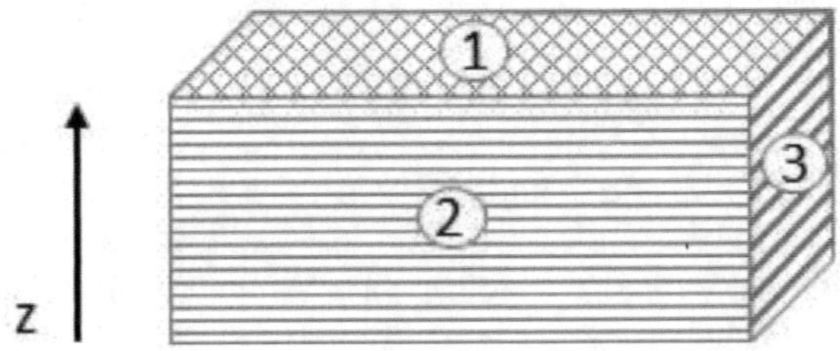

Figure 4. Planes of hardness measurement.

Thus, the average hardness of the original samples is 90-92 HRB, after HIP processing the average value decreased to 88-90 HRB, the results of measuring the hardness of surfaces 1-3 are at the same level. Also, as a result of HIP processing, there is a decrease in porosity, which is also reflected in [14, 15].

4. Conclusion

In this work, experimental data on the melting of stainless steel by the SLM method are obtained. The results of the study of the phase transformations of the material, the physical and mechanical characteristics of the parts obtained using the integrated technology by the SLM and HIP processes are reflected. Thus, as a result of the analysis of the porosity of a group of samples, the effect of HIP on a decrease in the quantity and size of pores, as well as on a decrease in hardness, was established.

Acknowledgements

This work was supported by the Russian Foundation for Basic Research (Grant 20-08-00310).

References

[1] Shishkovsky I V, Yadroitsev I A and Smurov I Y 2013 Manufacturing three-dimensional nickel titanium articles using layer-by-layer laser-melting technology *Tech. phys. letters* **39(12)** 1081-1084

[2] Smurov I Y, Movchan I A and Yadroitsev I A 2011 Laser-aided additive production *Vestn. MGTU Stankin* **2(4)** 144-146

[3] Tillmann, W, Schaak C, Nellesen J, Schaper M, Aydinöz M E and Hoyer K P 2017 Hot isostatic pressing of IN718 components manufactured by selective laser melting *Additive Manufact.* **13** 93-102

[4] Tarasova T V, Nazarov A P and Prokofev M V 2015 Effect of the regimes of selective laser melting on the structure and physicomechanical properties of cobalt-base superalloys *The Phys.of Metals and Metallography* **116(6)** 601-605

[5] Kamat A M and Pei Y 2019 An analytical method to predict and compensate for residual stress-induced deformation in overhanging regions of internal channels fabricated using powder bed fusion *Additive Manufact.* **29** 100796

[6] Leuders S, Thöne M, Riemer A, Niendorf T, Tröster T, Richard H A and Maier H J 2013 On the mechanical behaviour of titanium alloy TiAl6V4 manufactured by selective laser melting: fatigue resistance and crack growth performance *Int. J. Fat.*, **48** 300-307

[7] Lavery N P, Cherry J, Mehmood S, Davies H, Girling B, Sackett E and Sienz J 2017 Effects of hot isostatic pressing on the elastic modulus and tensile properties of 316L parts made by powder bed laser fusion *Materials Sci.and Eng.*: A **693** 186-213

[8] Sames W J, List F A, Pannala S, Dehoff R R and Babu S S 2016 The metallurgy and processing science of metal additive manufacturing *Int. Materials Review* 1–46

[9] Voznesenskaya A A, Kochuev, D A, Kireev A V, Zhdanov A V and Raznoschikov A S 2020 Optimization of laser melting parameters of martensity-aging steel powder *IOP Conf. Series: Materials Sci. and Eng.* **896(1)** 012132

[10] Boschetto A, Bottini L and Veniali F 2017 Roughness modeling of AlSi10Mg parts fabricated by selective laser melting *J. of Materials Proc. Tech.* **241** 154-163

[11] Kruth J P, Deckers J, Yasa E and Wauthlé R 2012 Assessing and comparing influencing factors of residual stresses in selective laser melting using a novel analysis method *Proc.of the institution of mech.eng., Part B: J. of Eng. Manufact.***226(6)** 980-991

[12] Usov S, Tochilin I, Voznesenskaya A and Zhdanov A 2019 Evaluation of microstructure and physical and mechanical properties of bulk parts obtained from Russian and German steel powder by selective laser melting *J. of Phys.: Conf. Series* **1331(1)** 012021

[13] Raznoschikov A S, Kochuev D A and Voznesenskaya A A 2020 Selective Laser Melting of Stainless Steel with Subsequent Processing *In Materials Sci. Forum* **989** 845-849

[14] Santorinaios M, Brooks W, Sutcliffe C J and Mines R A.W 2006 Crush behaviour of open cellular lattice structures manufactured using selective laser melting *WIT transactions on the built environment* **85** 481-490

[15] Gan M X and Wong C H 2016 Practical support structures for selective laser melting *J. of Materials Proc.Tech.* **238** 474-484

IOP Conf. Series: Materials Science and Engineering 1118 (2021) 012012 doi:10.1088/1757-899X/1118/1/012012

Stabilization of Root Parameters for Shielded Arc Welding

P P Krasikov, A V Savinov, O A Polesskiy, A A Chudin, L S Krasikova, I V Kozlov, D S Borisov and V V Filippov

Volgograd State Technical University, Lenin Av. 28, Volgograd 400005, Russia

E-mail: svarka-nv@vstu.ru

Abstract. The study revealed that the application of flux paste to the back side of the welded joint during welding of low-alloyed steels provides control over the sizes of the root and its defect-free formation. It was found that the back bead height of the root welded with flux paste is 1.5 to 3 times less than that of the root welded with the same welding variables without flux paste.

1. Introduction

Modern production rates require high and consistent quality of the welded joints.

In terms of economy and performance for many constructions the most efficient and sometimes the only possible mode is series welding with unsupported formation of the root. Such joints are the least laborious.

The study of reasons for the defects of these joints reveals unacceptable deviations in the root dimensions as the main problem which results in a poor penetration or a burn-through [1].

There is a number of techniques and methods being used to enhance the efficiency of welding and fabrication. The development of improved, extra proficient and cost-effective techniques is always targeted by researchers and industry [2-4].

The work [2] studied the modified short circuit GMAW process for the root seam and compared it with the conventional GMAW process. The comparison has been done on the parameters like current, voltage, wire feed rate, and travel speed and it was found that the weld quality in modified short circuit GMAW has been improved with the smooth appearance of the weld bead along with less spattering.

The literature has studied in sufficient detail welding process with single-component activating fluxes applied on the plate face, which increases penetration due to less surface tension of the weld pool [6-9, 11-14] and arc contraction [5,10].

The studies of automatic submerged arc welding also proved that there is an increase in tensile strength with increase in TiO_2, SiO_2 and CaF_2 content of coating mixture whereas microhardness significantly reduced with higher increase of SiO_2 content [15]. In previous research it has been studied that flux constituents and microstructure developing during welding; widely affect the mechanical behaviour of welds. In case of submerged arc welds the content of oxygen has considerable effect on microstructure or impact behaviour [16-18].

But there is almost no information on the flux pastes application to improve the root formation and its dimensions stabilization, which is the subject of this study.

2. Experimental part

Experiments in penetration of low-alloyed steel plates were performed with a non-consumable electrode in argon; the whole process was recorded with the video camera with optical filters (fig. 1). A number of refractory compounds were applied as components, such as SiO_2, ZrO_2, Al_2O_3, TiO_2, CaO, CrB_2, ZrB_2, TiB_2, CaF_2 with particle size up to 60 μm. They were dissolved in alcohol into suspension and applied on the back side of the plate with the coating thickness 200 - 250 μm. With the optimal composition of the flux paste selected, the 10-mm thick low-alloyed steel plates were welded automatically with a

Content from this work may be used under the terms of the Creative Commons Attribution 3.0 licence. Any further distribution of this work must maintain attribution to the author(s) and the title of the work, journal citation and DOI.
Published under licence by IOP Publishing Ltd

consumable electrode in the shielding gas mixture. Without the flux paste the chosen welding variables of automatic welding do not ensure consistent formation of the back bead.

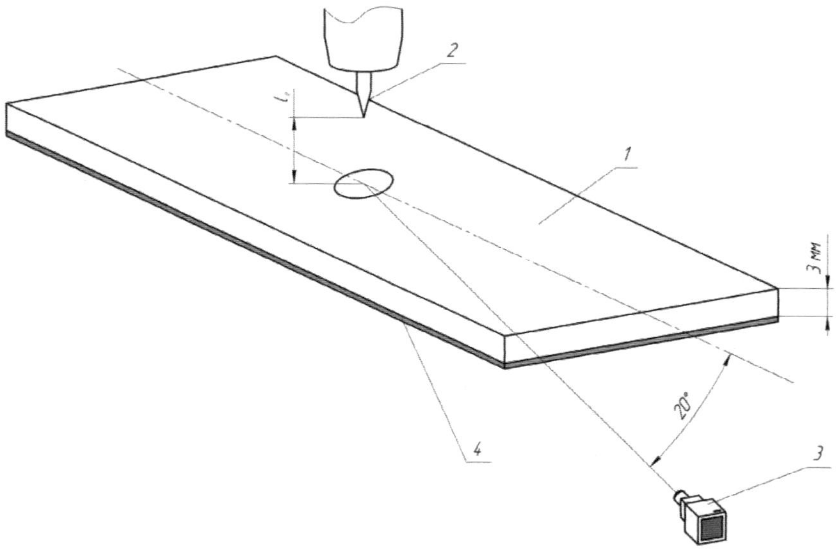

Figure 1. Scheme of the experiment: 1 – plate, material 09G2S; 2 – a non-consumable electrode $d_{el.}$ = 4 mm, $l_д$ = 2 mm; 3 –SAMSUNG VP-HMX-20C camera; 4 – coating from the selected compounds; penetration mode: $I_{w.}$ = 240 A, $t_{w.}$ = 4 s.

3. Results and discussion

The studies indicate that the beginning of the back bead formation without flux paste is characterized with fast increase in the back bead dimensions. The height increases from about 0.5 mm to 5.02 mm, and its width – from 4.7 to 10.7 mm in the time interval from 3.04 s to 3.26 s (in 0.22 s). The situation is different when the back bead is formed using zirconium oxide-based flux paste: the height increases from 0.46 mm to 3.08 mm, and the width - from 5.18 mm to 12 mm in the time interval from 3.26 s to 3.78 s (in 0.52 s). It proves that the applied coating can retain larger volume of the weld pool compared to the penetration without flux paste as well as protect from the negative influence of the arc and eliminate such root defects as a burn-through and deviations of geometric dimensions from permissible ones

The experiments with the selected components proved that the weld pool melt is retained when using all the selected active compounds. The most significant retention was observed with flux paste on the basis of aluminium and calcium oxide: the maximum sag height was 2.25 mm and 2.65 mm respectively (figure 2). Apparently, it is associated with the higher surface tension of the initial substances. For example, according to the reported data [19] the surface tension of Al_2O_3 = 580 mN/m, CaO = 510 mN/m, while of TiO_2 = 250 mN/m, SiO_2 = 280 mN/m.

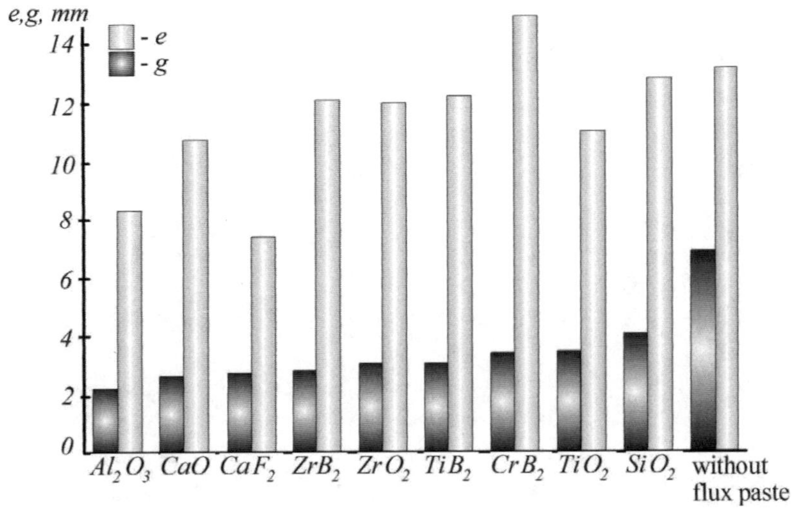

Figure 2. Dimensional characteristics of the back bead for different components applied: e – bead width; g – bead height

The study of root formation kinetics with the automatic welding with a consumable electrode in the shielding gas mixture without flux paste proved that at certain welding modes the weld pool sags with subsequent melt break-through out of the groove which results in burn-throughs (figure 3, a). When the flux paste is used as an active component (Al_2O_3) for the welding at the same modes, the weld pool sagging is acceptable and the formation of both face and back side of the root is defect-free (figure 3, b).

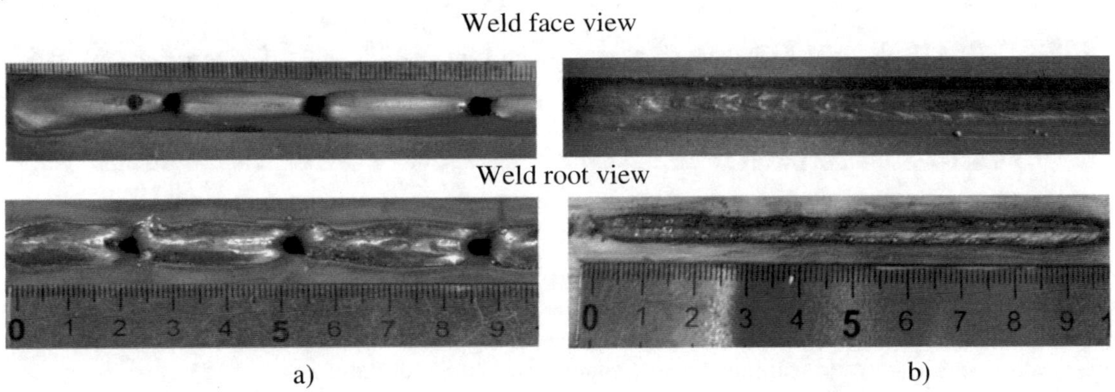

Figure 3. Weld root after welding with single-component Al_2O_3 flux paste (b), and without it (a): I_w = 270 A; U_{arc} = 29 V; V_w = 34 m/h; d_e = 1.2 mm; Vp.p. = 8 m/min; shielding gas 82%Ar+18%CO_2

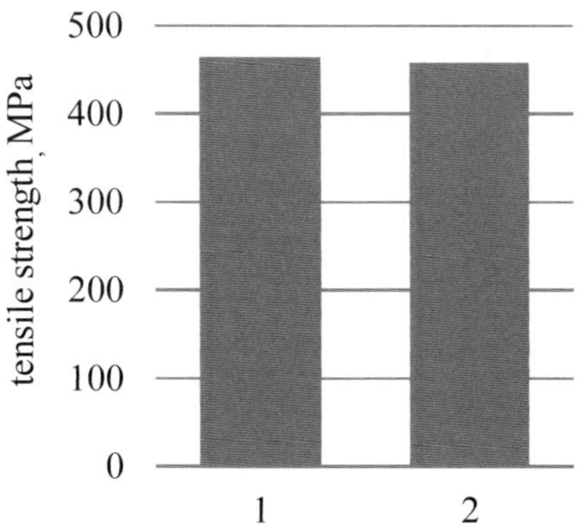

Figure 4. Tensile strength of the welded joints: 1 – without flux paste, 2 – with Al_2O_3

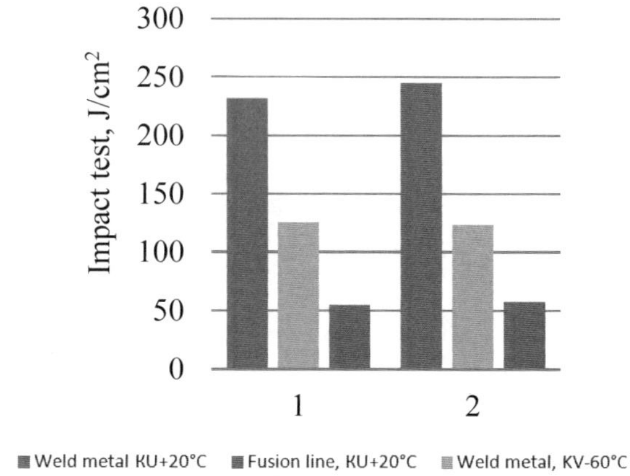

Figure 5. Results of impact tests of the welded joints: 1 – without flux paste, 2 – with Al_2O_3

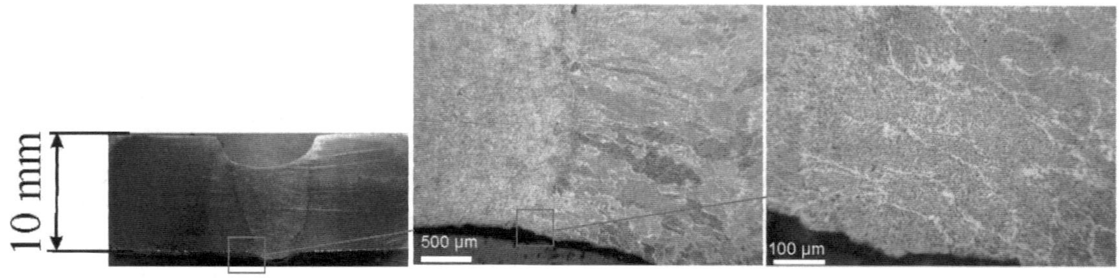

Figure 6. *Macro- and microsection of the resulted welded joint*

Mechanical tests of the welded joints proved that the tensile strength and the impact hardness for welding with coatings are on a par with the base metal for 09G2S steel (figure 4, 5). The macro- and microsections (figure 6) demonstrate the defect-free root with deeper penetration, which allows to use high-performance automatic welding for subsequent filler bead formation.

4. Conclusion

Applying the flux paste on the back side of the welded joint during welding low-alloyed steels makes it possible to control the sizes of the root and its defect-free formation. The back bead height of the root welded with flux paste is 1.5 to 3 times less than that of the root welded with the same welding variables without flux paste.

References

[1] Krasikov P P, Savinov A V, Polesskiy O A et al 2020 Formation of a seam using automatic unsupported argon arc welding with a non-consumable electrode with flux pastes *J. Phys.: Conf. Ser.* **1443** 012014
[2] Singaravelu D L, Rajamurugan G and Devakumaran K 2018 Modified short arc gas metal arcwelding process for root pass welding applications. *Mater Today Proc* **5(2)** 7828–35
[3] Norrish J and Cuiuri D 2014 *J. Manuf. Process.* **16(1)** 86-92
[4] Mohammad Mousavi Anzehaee, Mohammad Haeri 2012 *J.Process Contr.* **22(6)** 1087-1102

[5] Kumar V, Lucas B, Howse D, et al 2009 *JOM-15- Fiftheenth International Conference on the Joining of Materials, Denmark, Helsingor*, pp 1–11
[6] W.X.-h. L. Qing-ming, Z. Zeng-d, W. 2007 *Jun, Trans. Non-ferrous Met. Soc. China*, **17** 486-490.
[7] Vora J J, Badheka V J, 2015 *J. Manuf. Processes* **20** 224–233
[8] Vora J J, Badheka V J, 2016 *Trans. Indian Inst. Met.* **69** 1755–64
[9] Lu H F S, H. Sugiyama, Tanaka M, Nogi K, 2002 *Mater. Trans.* **43** 2926–31
[10] Vora J J, Badheka V J 2019 Insights into the flux-assisted TIG welding processes, in: (Eds.), Advances in Welding Technologies for Process Development (CRC Press, Taylor and Francis)
[11] Mills B J K K C, Brooks R F, Shirali A, 1998 Phil. *Trans. R. Soc. Lond., A* **356** 911-925
[12] Berthier A, Paillard P, Carin M et al 2012 *Sci. Technol. Weld. Joining* **17** 616–621
[13] Berthier A, Paillard P, Carin M et al 2012 *Sci. Technol. Weld. Joining* **17** 609–615
[14] Bhattacharya A 2016 *Mater. Manuf. Processes* **31** 343–351
[15] Bhandari D, Chhibber R, Arora R, Mehta R 2016 Investigation of TiO2–SiO2–CaO–CaF2 based electrode coatings on weld metal chemistry and mechanical behaviour of bimetallic welds *J. Manuf. Process.* **23** 61-74
[16] Golovko V V, Potapov N N 2011 Special features of agglomerated (ceramic) fluxes in welding *Weld. Int*. **25** 889-893
[17] Maalekian M 2007 The Effects of Alloying Elements on Steels (I) (Technische Universität Graz, Austria)
[18] Kohno R, Takami T, Mori N, Nagano K 1982 New fluxes of improved weld metal toughness for HSLA steels *Weld. J.* 6**1 (12)** 373-380
[19] Podgaetsky V V and Kuzmenko V G 1988 Welding slag. Handbook (Kyev: Naukova Dumka) in Russian

Cold deformation of non-circular drill pipes

S V Parshin[1] and A I Dronov[2]

[1]Ural Federal University named after the first President of Russia B. N. Yeltsin, 620000, Mira st., 19, Yekaterinburg, Russia
[2]JSC Pervouralsk Novotrubny plant, Torgovaya st., 1, Pervouralsk, Sverdlovsk region 623112, Russia

E-mail: netskater@mail.ru

Abstract. In this paper, the pipe billet dimensions are determined according to the required kelly configuration. A finite element model of the deformation center at drawing junctions is created. Based on the stress-strain calculations, the following invariants are determined: hydrostatic pressure, load, and degree of shear strain. The metal damage extent is determined and proved to be reducible by performing the two-pass drawing. It is shown that the 'pendulum' heat treatment provides for complete healing of the metal defects caused by plastic deformation. As a result, a technology of producing high quality kellies is proposed.

1. Introduction

Kellies are an integral part of drill strings that are designed to transfer torque to the working tool. They largely determine the penetration efficiency of oil and gas wells. The main part of a kelly is a non-circular thick-walled tubing with an internal channel [1]. The most effective way to produce such tubings is to draw them in a profile die on a moving mandrel. However, to date, there is no theoretical justification for this plastic deformation process. To provide it, a number of problems must be solved, including determination of the billet size and shape, the longitudinal and transverse configuration of the working tool, the number of broaches, extent of the metal damage and its healing capacity under heat treatment.

2. Methods

The analysis has revealed that it is necessary to combine centers of gravity of the billet and the finished profile and to ensure equality of the cross-section drawing of the profile elements [2]. For this purpose, a round or profile billet, similar to a finished profile, can be used. In the current study, a round thick-walled tube was used as a billet. The billet wall diameter and thickness were calculated based on the finished profile dimensions.

Conditions of the studied process had to ensure the required quality of the pipe metal. For this purpose, extent of the metal damage ω_0 was measured and the corresponding heat treatment for restoration of the metal structure after plastic deformation was determined.

The billet wall diameter and thickness were determined based on the research described in [3]. Then, the principle of least action was applied. This made it possible to determine the metal flow lines [4, 5] that were approximated using cubic splines [6]. Additionally, equipotential lines were drawn to determine the die output cross-section at transitions [7, 8] (Figure 1).

Content from this work may be used under the terms of the Creative Commons Attribution 3.0 licence. Any further distribution of this work must maintain attribution to the author(s) and the title of the work, journal citation and DOI.
Published under licence by IOP Publishing Ltd

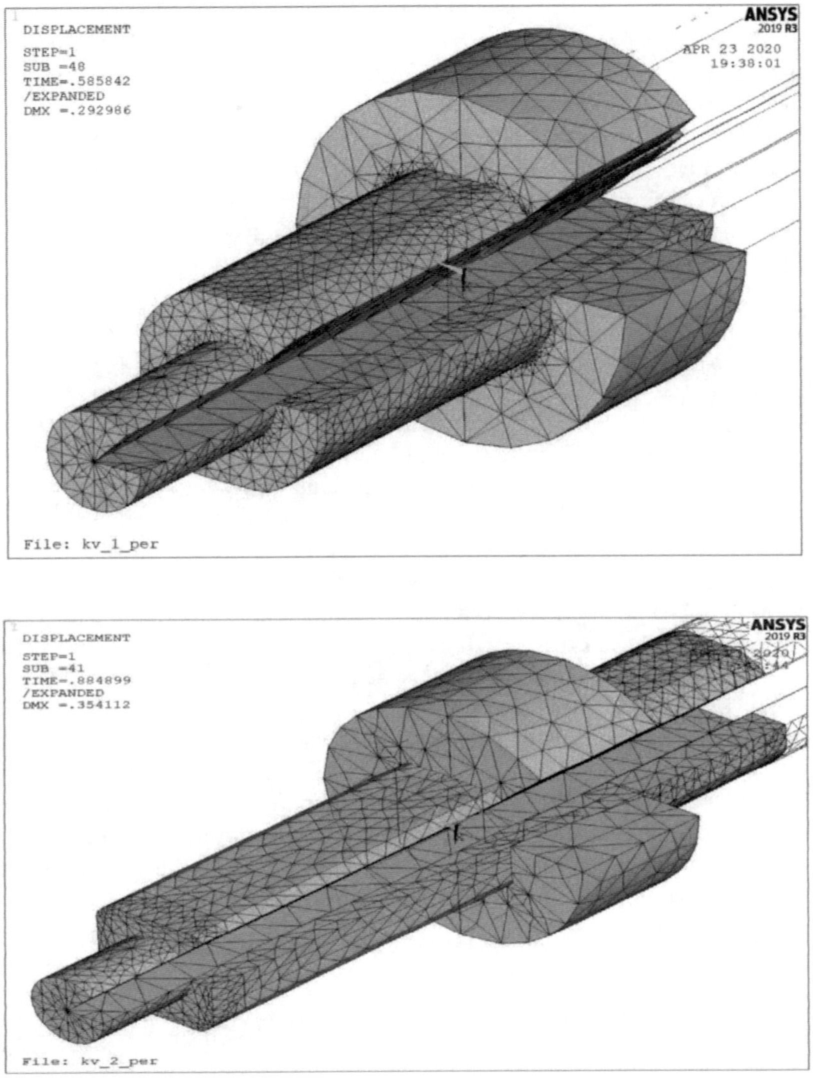

Figure 1. Finite element model of the deformation center: (a) and (b) are the first and the second deformation passes, respectively.

As a result, a model of producing square pipes in two passes by drawing them on a long moving mandrel was created. The solid model included three bodies: a smooth billet model, a long cylindrical mandrel, and a profile die. To reduce the processing time, we considered the problem from the plane symmetric perspective (1/8 of a full profile). The solid model was exported in the SAT format (developed by Spatial Corp.).

After transfer to the finite element package, the model was split using elements of the tetrahedron type as they are the most suitable for pipes of complex curvilinear shapes. The grid density on contacting surfaces of the mandrel, die, and the pipe billet was additionally increased. This made it possible to determine the finished profile shape and to make the surfaces more 'smooth' and transfer them with a better precision, which reduced the probability of errors at modelling of the relative slip on contacting surfaces.

Two sets of contacting surfaces were considered: that of the mandrel with the pipe model inner surface and of the die working surfaces with the pipe outer surface. The coefficient of friction was set according to the tabular data taking into account in-situ measurement values. The friction coefficient was adjusted based on actual values of the energy and power parameters (in particular, the drawing force). The mandrel and the pipe movement ranges were set. The die was set to be fixed and nondeformable. The symmetry planes were also specified.

The material properties were set as a multi-linear hardening curve that factored in both elastic and plastic properties of the pipe material. The finite element package allowed for taking into account instantaneous variable properties of the material of different model elements. This made it possible to take into account the elastic effect of individual profile elements after their leaving of the deformation zone, which had a positive impact on the geometric modeling accuracy.

After determining the pipe shape at the first transition, the profile was approximated using arcs and segments. Then, the resulting polyline was used as a basis for modeling geometry of the pipe at the second transition. The transition damage measured at reference points of the profile was summed up to determine the full extent of the metal damage.

The solution was to ensure the intermediate transition rational shape in order to reduce the working tool wear, minimize the pipe material damage (by increasing the deformation fractionality), and produce non-circular pipes meeting the specification requirements. This was achieved due to fuller filling of the profile corners and consequent reduction of the rounding radii at the profile edges.

The application of the finite elements method made it possible to determine all invariants of the metal stressed and deformed conditions and its damage (ω0) at 'dangerous' points of the pipe cross-sections at each deformation stage. As known, there exist critical values (ω^* and ω^{**}) that should be compared with ω_0. Within the range of these critical values, defects caused by plastic deformation are healed differently. Thus, if $0<\omega_0<\omega^*$, defects are completely healed during further annealing. If $\omega^*<\omega_0<\omega^{**}$, only some of the defects are healed [10, 11, 12].

At the first stage, the single-pass drawing was studied. It was found that on the profile edges $\omega_0>\omega^{**}$, i.e. more than the allowable value. To increase the deformation fractionality and reduce ω_0, the two-pass drawing was applied. At the first transition $\omega_0=0.69$, at the second - 0.534, which, in total, also exceeded the allowable value. For this reason, in between the passes, the 'pendulum' annealing should be applied. It allows to expand the boundaries of reversible damage from 0.37 to 0.5 by increasing the size of the healed defects. The healing value was calculated using the following formula [9]:

$$\Delta\omega = \omega^* + \frac{c_\zeta - \omega^*}{\omega^{**} - \omega^*}(\omega_0 - \omega^*) \quad (1)$$

The resulting residual damage for the first pass is $\omega_{r1}=0.14$, for the second - $\omega_{r2}=0.126$, which is acceptable.

3. Conclusion

It was found that the preferable way of kelly production includes two drawing passes with 'pendulum' annealing after each of them.

This method proved to provide the following benefits:

1. Obtaining a precisely shaped profile due to calibration and elimination of temperature distortions during the heat treatment at the last transition.

2. Acceptable values of material damage due to its significant reduction during the heat treatment.

Besides, repetitive nature of the heat treatment makes it possible to control the product properties, e.g. using express methods of the surface hardness measurement, which prevents failures related to mechanical properties.

This approach provided the best results in terms of quality and reliability of finished products.

References

[1] Pattillo P D 2018 *Elements of Oil and Gas Well Tubular Design* (Elsevier: New York)
[2] Vydrin V N, Grosman A B, Pavlov V K and Nagornov V S 1977 Production of High-precision Shaped Profiles (Moscow: Metallurgia) [in Russian]

IOP Conf. Series: Materials Science and Engineering 1118 (2021) 012013 doi:10.1088/1757-899X/1118/1/012013

[3] Danchenko V N, Sergeev V V and Nikulin Je V 2003 Production of Shaped Pipes (Moscow: Intermet Engineering) [in Russian]
[4] Gun G Ja 1980 Theoretical Foundations of Metal Forming (Moscow: Metallurgia)
[5] Gun G Ja 1983 Mathematical Modeling of Metal Forming Processes (Moscow: Metallurgia)
[6] Ahlberg J H, Nilson E N and Walsh J L 1967 *The Theory of Splines and Their Applications* (New York: Academic Press)
[7] Parshina A A 2017 Determination of the profile of the working tool for drawing pipes with variable wall thickness *Proizvodstvo prokata* **3** 36 [in Russian]
[8] Parshina A A and Parshin S V 2013 Computer simulation of the pressure treatment process when obtaining shaped pipes with variable wall thickness/ *IMMB: UrFU*, **1** 93 [in Russian]
[9] Smirnov S V and Shvejkin V P 2009 Plasticity and deformability of carbon steels during pressure treatment (Ekaterinburg: UGTU-UPI) [in Russian]
[10] Bogatov A A, Mizhirickij O I and Smirnov S V 1984 *Resurs plastichnosti pri obrabotke metallov davleniem* (Resource of plasticity in metal forming) (Moscow: Metallurgia)
[11] Lemaitre J 1991 *Handbook of Materials Behavior Models* (Dordrecht: Springer Netherlands)
[12] Smirnov S V, Bogatov A A and Kolmogorov V L 1980 Investigation of plastic loosening of metal and healing of deformation defects during annealing *Fizika metallov i metallovedenie* B **49** 389 [in Russian]

IOP Conf. Series: Materials Science and Engineering 1118 (2021) 012014 doi:10.1088/1757-899X/1118/1/012014

iLabit Robotica: A Photo-Realistic Extension for Robotics Toolbox

Ivan Kholodilin[1,2], Nikita Savosteenko[2], Yuri Abramov[2] and Luis A Lago[1]
[1] Beijing Institute of Technology, Beijing 100081, China
[2] South Ural State University, Chelyabinsk 454080, Russia

E-mail: kholodilin@bit.edu.cn

Abstract. This paper presents a photo-realistic extension for Robotics Toolbox (iLabit Robotica). Built using Unity, the proposed simulator integrates the 6-axis robot arm, sensors, and elements for interactions. iLabit Robotica is aimed to increase the quality of the engineering education and motivation of students towards the educational process, by providing a new beginning for the project-based learning (PBL) approach with the aid of virtual technologies. We strongly believe that by simulating real technological processes and integrating them into the educational process, students will not only achieve a valuable practical experience by solving "real-life" existing problems, but also a substantial interest towards the subject. We demonstrate the compatibility between our simulator and Robotics Toolbox by interacting with objects placed inside the virtual environment through Matlab via TCP/IP. Experiments demonstrate that the operating process is simple and it provides robust results and a higher measurement accuracy. The proposed simulator and supporting materials are available online: http://www.ilabit.org.

1. Introduction

Specialized software products capable of providing three-dimensional visualization, e.g. a technological process or high-precision mathematical modelling such as Robot Studio, SOLIDWORKS, and ANSYS, were developed and widely used at the end of the last century. This kind of software is implemented in almost all segments of the engineering industry, ranging from the defence industry, heavy industry, and aerospace engineering to microelectronics, medicine, and simulators. Modelling and analysis in several industries may avoid costly and time-consuming design-build-test cycles. As a result, the widespread use of different types of software in industry, educational institutions started to integrate these software products into the educational process. Results showed that teaching courses based on specialized software products had a significant impact on the level of knowledge acquisition, competencies, and even motivation of science and engineering students [1]. The research on the quality of education received by students in higher educational institutions shows that the introduction of specialized software had a significant impact on the level of education and competencies of graduates, especially in engineering-related specialities [2].

Nowadays one of the fastest growing areas in engineering is industrial robotics, namely robotic manipulators. Robot manipulators are complex and expensive technical objects that have found wide applications in medical [3], and industrial fields [4]. In order to work with robot manipulators specific programming skills are required, understanding of the technological process based on these robots, as well as deep knowledge of the theory of electric drives and automation. However, sometimes it can be

Content from this work may be used under the terms of the Creative Commons Attribution 3.0 licence. Any further distribution of this work must maintain attribution to the author(s) and the title of the work, journal citation and DOI.
Published under licence by IOP Publishing Ltd

difficult or even impossible to recreate real technological processes in a classroom environment. Moreover, in many cases the resources used as a necessary means to successfully carry out some of these activities are very expensive, e.g., industrial robots. At the same time, project-based learning (PBL) approach is an important tool that can be successfully applied in a variety of scenarios. It has proven to be effective in different situations, while increasing the interest of students towards them not only in higher education [5], but in the middle school education as well [6]. Therefore, the finding of new interesting and cost-effective ways to implement PBL in engineering education remains as a relevant topic. Secondly, during the 2020 year we have witnessed how the educational process has been interrupted to some extent. Engineering students have required laboratories to test their theories and conduct practical experiments and they have lost this possibility given this difficult situation. Therefore, new ideas should be explored in online education. Last but not least, for college graduates it can take a long time to get accustomed to the industrial processes, which, at the same time, can be very time- and cost-consuming for employers. Consequently, new ways of interactions are required on early stages (before graduation from universities), aimed at preparing students to the “real life” in a variety of working environments. It is worthwhile to note that it would be better if these new methods do not distract students from the main study.

By taking the aforementioned problems into account, we can state that the issue of developing a specialized virtual laboratory, which will provide the possibility to carry out a wide range of experiments, is of particular importance today. One of the possible existing solutions is the one developed by ABB “Robot Studio” software product, which not only supports a wide range of scenarios, but also allows to interact with industrial robots, and at the same time, immerse the user inside the technological process through realistic three-dimensional graphics. The main disadvantage of this software is that users do not have the ability to entirely interact with an industrial robot, namely with electric drives belonging to the corresponding joints of the robot. This drawback can be eliminated with the use of the “Robotics Toolbox” software package for MATLAB [7]. The Robotics Toolbox is a software package that enables MATLAB user to readily create and manipulate datatypes fundamental to robotics such as homogeneous transformations, quaternions, and trajectories. Besides of the programming algorithms for manipulating with the robot manipulator, this Toolbox allows to use mathematical models of electric drives. The limitation of this program is the lack of realistic graphics and interaction with objects. However, by using modern game engine platforms, e.g. Unity, it is possible not only to visualize the technological process, but also to integrate physics into it. In the view of the current problems and trends in engineering education, this paper proposes a photo-realistic virtual laboratory, which includes a robot manipulator fully compatible with the Robotics Toolbox.

2. Simulator overview

There are existing robotics toolboxes [7-11] which enable students and teachers to better understand the theoretical concepts behind classical robotics and computer vision through easy and intuitive simulation and visualization. It is worthwhile mentioning that the most popular one was developed by P. Corke and called “Robotics Toolbox” [7]. The toolbox is based on a very general method of representing the kinematics and dynamics of serial-link manipulators using objects. Each link is represented by a link object whose attributes are the standard or modified Denavit and Hartenberg (DH) parameters, as well as the link and motor inertial properties, friction, and gear ratio. Functions in this toolbox are written in a textbook manner, which helps to better understand theory by practice. However, the Robotics Toolbox has particular limitations as well. For example, Matlab environment is not really suitable for visualization tasks as it suffers from the photorealism. In contrast, more flexibility towards visualization can be provided by game engines such as Unity, UNIGINE, CRYENGINE, and Unreal Engine 4. Moreover, it was found earlier that virtual technologies are more attractive for students than traditional educational materials [2]. By taking this information into consideration we decided to provide new insights towards engineering education by creating an extension for the Robotics Toolbox. Our simulator is built by Unity, the main screen of the simulator is presented in Fig. 1. Main parts of the scene consist of the industrial robot and the conveyer with colored objects. The manipulator includes the camera attached to the end effector (EE); this camera can be activated/hidden by the corresponding button located on the panel. Next to the camera button is located the button that activates the coordinates of the joints of the manipulator. This information can be useful for describing the manipulator in terms of DH parameters [12] or modified DH parameters [13].

Figure 1. Main screen of the proposed simulator. The left blocks of the menu are in charge of the interaction with the manipulator and objects on the conveyer; the right block shows the streaming of the camera attached to the end effector.

The proposed simulator can be simply installed and configured on Windows, macOS, and Linux operating systems. The simulator provides three main features: (1) interaction with scenes and objects included into the virtual environment, (2) communication with other programs by TCP/IP, (3) a detailed user guide and technical support. All these features allow to configure experimental conditions for testing theories and improving operating algorithms of the system. During the development of the simulator, a particular attention was dedicated to creating photorealistic scenes, by using rendering capabilities such as light sources, reflection from the surfaces, shadows, and so on. Fig. 1 shows a snapshot taken by our simulator which reveals rendering capabilities. The simulator supports two modes: manual mode and through communication with other programs by TCP/IP (see Table 1).

Table 1. Operating modes of the simulator.

	Manual mode from the user interface (UI) of the simulator	Communication with other programs by TCP/IP
Interaction with the robot	The angles of the joints can be controlled by sliders or input fields (see Fig. 1)	The angles of the joints can be sent through TCP/IP separately, e.g. forward kinematics or in a loop, e.g. inverse kinematics.
Interaction with the objects[a]	EE can grab and release objects by toggle "Grab".	EE can grab and release objects by discrete parameter "Grab".
Activating new objects	Generating "Green Cube", "Red Cube", "Random Cube" or removing Cube from the conveyer with the corresponding buttons (see Fig. 1).	Generating "Green Cube", "Red Cube" or "Random Cube" or removing Cube from the conveyer with the corresponding combination of bits.

[a] The distance between the EE and the object can be controlled from zero (for precise operating tasks) to bigger values (when high precision operating is not required). For example, if this distance is set up equal to zero and there is any misalignment between the EE and the object, consequently, in this particular case, it is not possible to grab the object.

3. Experiment

We strongly believe that the proposed simulator is able to bring new insights towards robotics course by interacting with objects and modern graphics. Thus, the main goal of the experiment was to investigate the compatibility between the Robotics Toolbox in Matlab and our simulator. It was decided to demonstrate this compatibility by means of practical tasks, namely by sorting six objects (three green and three red) with the known coordinates of them and joints of the robot manipulator. The simulation model of the robot was based on the real parameters of the manipulator Motoman MH50II (see Fig. 2). More details about the experiment are provided bellow.

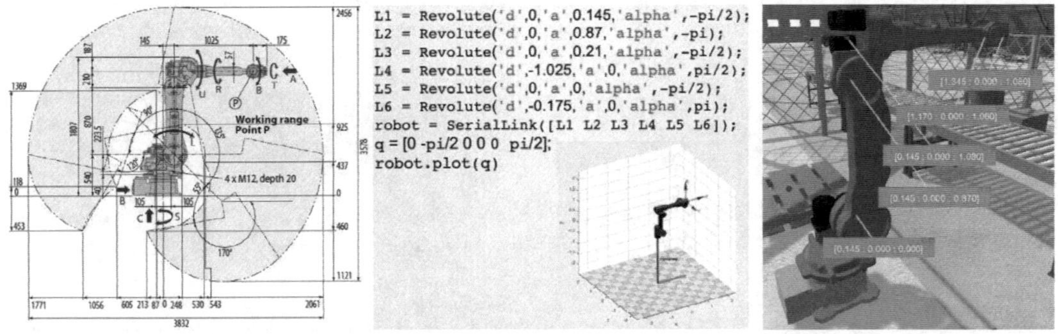

Figure 2. Simulation of the manipulator. From left to right: parameters of the manipulator, model built by Robotics Toolbox, model inside the simulator.

3.1. Experiment setup

The proposed simulator is able to interact with objects included into the virtual environment, such as objects on the conveyer and pallets. Consequently, in order to operate with these objects through Matlab it was necessary to provide an understanding of their corresponding positions and sizes. Once the information about the objects is entered into the simulator it can be verified by carrying out the experiment with Matlab. This experiment was based on sorting the green and red objects to the corresponding pallets from the initial position on the conveyer with the coordinates (1.733; 0.000; 0.369). Red objects were sorted onto the centre of the right pallet. Calculations were based on given coordinates for the center of the right pallet (0.000; 1.350; -0.400), and the height of the objects was set at 0.158 meters. Green objects were sorted onto the left pallet in a more complicated way, namely to the corner. Calculations were based on given coordinates for the center of the right pallet (0.000; -1.350; -0.400), width and height of the pallet (1.015; 1.220), and the size of objects was set at (0.400; 0.400; 0.158). The above-mentioned calculations are depicted for the first objects in Fig. 3, the remaining objects should be placed above the corresponding green and red objects.

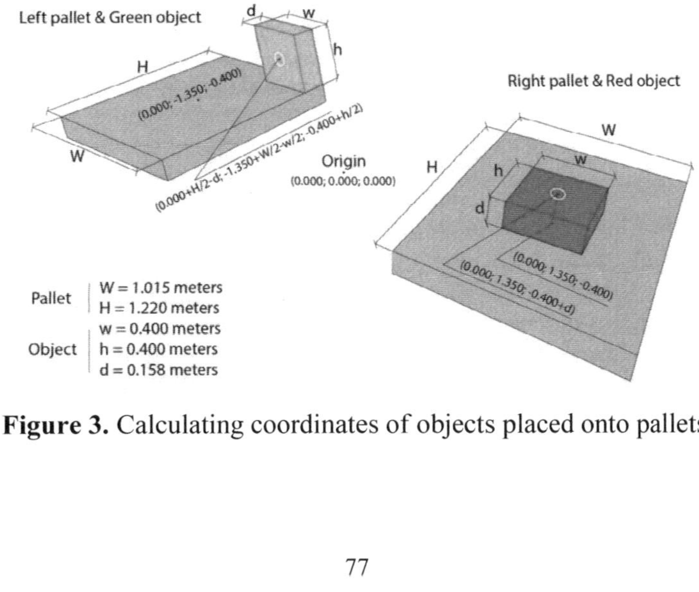

Figure 3. Calculating coordinates of objects placed onto pallets.

Finally, once all of the coordinates computed, the calculated Robotics Toolbox function "ikine" was used, which calculates inverse kinematics based on the given points and function "jtraj" calculating trajectory between given points. Angles of every joint were transmitted to the simulator through TCP/IP. Results are presented in the next subsection.

3.2. Results

Visual results of the experiment are shown in Fig. 4. From this figure it can be seen that the objects take a desired location without any visual misalignments between objects. More accurate comparison analysis is presented in Table 2. From the values in Table 2 it can be seen that the coordinates of the experiment are similar to the coordinates from the experiment setup, the difference between them is very small, in the range of a few millimetres.

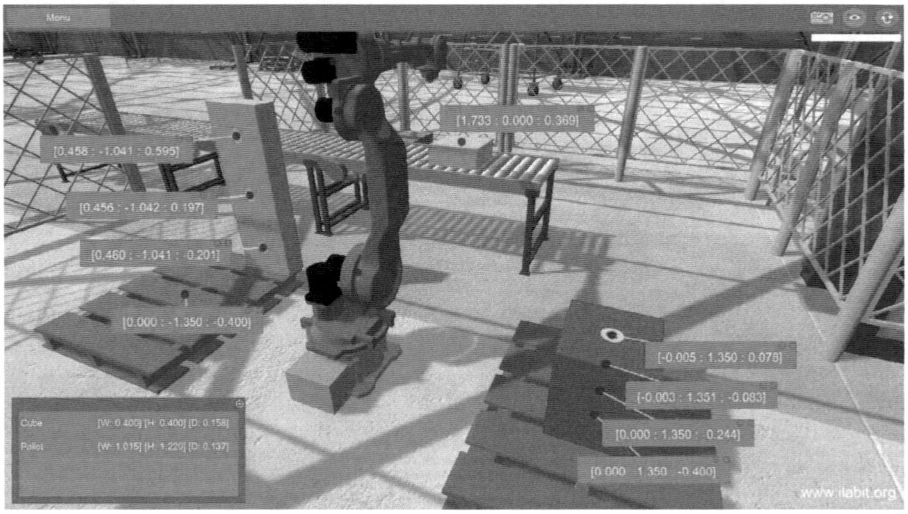

Figure 4. Results of the experiment.

Table 2. Comparison analysis.

Objects	No.	Calculated coordinates, meters	Experiment coordinates, meters	Absolute error, meters
	1st	(0.000; 1.350; -0.242)	(0.000; 1.350; -0.244)	(0.000; 0.000; 0.002)
Red	2nd	(0.000; 1.350; -0.84)	(0.000; 1.351; -0.83)	(0.000; 0.001; 0.001)
	3rd	(0.000; 1.350; 0.740)	(0.005; 1.350; 0.780)	(0.005; 0.000; 0.002)
	1st	(0,452; -1.042; -0.202)	(0,460; -1.041; -0.201)	(0.008; 0.001; 0.001)
Green	2nd	(0,452; -1.042; 0.202)	(0,456; -1.042; 0.197)	(0.004; 0.000; 0.005)
	3rd	(0,452; -1.042; 0.602)	(0,458; -1.041; 0.595)	(0.006; 0.001; 0.007)

4. Conclusion

The paper has demonstrated a publicly available extension for the Robotics Toolbox in MATLAB, the principle features, and its capabilities. The software product presented in this article showed the possibility of implementing virtual laboratories for studying robot manipulators, whereas the experiment results demonstrated its full compatibility with the Robotics Toolbox. By simulating real technological processes, it was possible to determine new ways for solving the problems defined earlier. We strongly believe that virtual technologies will not only influence on the motivation of students, but at the same time, it will save economic resources for educational centers, as virtual complexes are substantially less costly and the resources involved in the maintenance process are less demanding than those required by the real ones. The proposed simulator provides an opportunity to conduct experiments from any place, only a computer is required. Thus, even in a difficult situation students are able to combine theory with practice, which is of particular importance to each of them in their training as future engineers.

The literature review related with the process of creating virtual laboratories showed that this direction is young and is just emerging. In fact, in order to facilitate teaching and learning of the fundamental concepts in the subject, pedagogy of teaching the subject should take into consideration the latest technological tools and move with the times. An example of how the robotics course can be improved was shown in this paper. In the future, we aim to further work on the improvements of the simulator by including new scenarios and also consider new directions in which virtual laboratories can take place.

Acknowledgments

The work was supported by Foundation for Assistance to Small Innovative Enterprises, contract No. 477GUCES8-D3/61968 from 05.10.2020.

References

[1] Rodrigues C S and Rocha A M 2014 Analysing Students' Attitudes Towards the Learning of Specialized Software *Proc. Int. Conf. on Computational Science and Its Applications (Guimaraes)* 264-267

[2] Bazarov S E, Kholodilin I Y, Nesterov A S and Sokhina A V 2017 Applying Augmented Reality in practical classes for engineering students *Proc. of the IOP Conf. Ser.: Earth Environ. Sci. 87*

[3] Kucuk S and Gungor B D 2016 Inverse kinematics solution of a new hybrid robot manipulator proposed for medical purposes *Medical Technologies National Congress (Antalya)* p 1-4

[4] Zeng Q, Liu X, Qiu C and Li A 2019 Kinematic Characteristics Analysis of Cooperative Welding Robot with Multiple Manipulators 2019 *Proc. Int. Conf. on Intelligent Robot Systems (Nagoya, Japan)* p 1-5

[5] Mones A, Sanchez E, Dimitriadis Y A, Abellan I M, Avi B and Gorgojo G 2005 Multiple Case Studies to Enhance Project-Based Learning in a Computer Architecture Course *J IEEE Transactions on Education* **48** 482-489

[6] Timothy D W 2017 *Teach Like Finland* (W.W. Norton, Incorporated)

[7] Corke P I 1996 A Robotics Toolbox for Matlab *IEEE Robotics and Automation Magazine*.

[8] Vila-Rosado D N and Dominguez-Lopez J A 2005 A Matlab toolbox for robotic manipulators *Proc. Int. Conf. on Computer Science (Puebla, Mexico)* p 256-263

[9] Righettini P, Strada R, Olama E K and Valilou S 2015 Symbolic kinematic and dynamic modelling toolbox for Multi-DOF robotic manipulators *Proc. Int. Conf. on Automation and Computing (Glasgow)* p 1-7

[10] Elshatarat H L, Baniyounis M and Biesenbach R 2016 Design and implementation of a RoBO-2L MATLAB toolbox for a motion control of a robotic manipulator *Proc. Int. Conf. on Systems, Signals & Devices (Leipzig)* p 89-95

[11] Liu J and Luo Q 2019 Modeling and Simulation of Robotic Arm in MATLAB for Industrial Applications *Proc. Int. Conf. on Intelligent Human-Machine Systems and Cybernetics (Hangzhou, China)* p 346-349

[12] Paul R P 1981 *Robot Manipulators: Mathematics, Programming, and Control* (Cambridge, Massachusetts: MIT Press)

[13] Craig J 1986 *Introduction to Robotics* (Addison Wesley)

Estimation of electric loads dynamics in the gas fields of Western Siberia taking into account the life cycle of the fields

I M Bogachkov[1], R N Khamitov[2,3] and V P Frayshteter[3]
[1]Tyumen branch of LLC "Gazprom engineering", Tyumen, Russian Federation
[2]Omsk State Technical University, Omsk, 644000, Russian Federation
[3]Tyumen Industrial University, Tyumen, 625000, Russian Federation

E-mail: bogim83@mail.ru

Abstract: All gas fields go through four periods of their life cycle with their technological scheme changes. The engineering of electricity supply system for gas fields in West Siberia was performed without taking into account the final stage of their life cycle. To extend the period of profitability of low-pressure Cenomanian gas at the final stage of the life cycle an energy-intensive technology of gas compression distribution with usage of mobile compressor units (MCU) is implemented. The implementation of this technology in the gas fields requires the construction of a new electricity supply system or the reconstruction of the entire existing one. The objective of the paper is to assess the dynamics of electric loads of gas fields, considering the entire life cycle. The average annual calculated electric load for each life cycle period for active fields in West Siberia for 2019 is determined. The coefficient of increase in electric load is calculated. The tasks are solved using the empirical method (measurement, comparison) and the experimental-theoretical method (analysis and synthesis). The analysis shows that during the life cycle of a gas field, when the well outcrop and average reservoir pressure decrease over time, the calculated electrical load increases in many times.

1. Introduction

Currently, most of the major gas fields in Western Siberia are at the final stage of development of the Cenomanian Deposit.

To maintain gas production volumes at this stage, modern distributed gas compression technology is being implemented with the installation of mobile compressor units (MCUs) with an electric drive with a capacity of up to 1 MV at the mouth of gas well clusters.

In 80 - 90s, gas fields in Western Siberia were designed without taking into account electrical loads at the final stage of the life cycle, so now there are problems associated with the need for a major reconstruction of the power supply system, which significantly affect the level of profitability of gas production.

Thus, there is a need to assess the electrical loads in each stage of the life cycle at the existing gas fields in Western Siberia.

When designing, the electrical loads should be determined for a long period of field development because on the basis of these the power supply scheme will be formed and gradually developed. In the literature [1-7], the regularities of formation and development of electric loads of oil fields in Western Siberia are studied in sufficient depth and detail, and methods for calculating and predicting electric loads of oil production facilities are developed. But with the external similarity of production

Content from this work may be used under the terms of the Creative Commons Attribution 3.0 licence. Any further distribution of this work must maintain attribution to the author(s) and the title of the work, journal citation and DOI.
Published under licence by IOP Publishing Ltd

IOP Conf. Series: Materials Science and Engineering 1118 (2021) 012015 doi:10.1088/1757-899X/1118/1/012015

technology (extraction of hydrocarbon resources from wells drilled over a large area and their external transport), the patterns of formation and development of electric loads of oil and gas fields are fundamentally different.

2. Problem statement

In the process of developing of gas fields, there are several technological stages ("life cycles") characterized by different ratios of electricity costs per unit of gas production. This analysis is necessary to assess the dynamics of electrical loads over a long period, and when designing the development of a field, it is essential to determine the estimated electrical loads in each period of its life cycle.

The electrical loads based on which the power supply scheme will be formed and gradually developed should be determined for a long period of field development.

In the process of gas fields developing, several technological stages ("life cycles") can be distinguished, characterized by different ratios of electricity costs per unit of gas production. This analysis is necessary to assess the dynamics of electrical loads over a long period, and when designing the development of a field, it is necessary to determine the estimated electrical loads in each period of its life cycle. The main technological stages of gas field development, which are fundamentally different from the point of view of specific electricity costs, are as follows:

Stage of growing gas production. A relatively short period, characterized by growing, but small electrical loads and their minimum level per unit of gas production. The main consumers of electricity at this stage are complex gas preparation plants (CGPP), electric receivers of field gas pipelines (ERFP).

Stage of stabilization of gas production at the maximum (with a slight decrease) level. The electrical loads of this stage are stable, and booster compressor stations (BCS) are added to the electric receivers of the first stage. The cost of electricity per unit of gas production at this stage is also low for the entire period of stable gas production.

The next stage of the field's life cycle, the longest one, is the stage of falling gas production. When the level of electrical loads is almost unchanged from the previous technological stage, as a result of an intensive decrease in gas production volumes, the unit cost of electricity per unit of gas production increases sharply.

To intensify gas production at fields that are at the final stage of development, distributed compression technology has recently been used using small-sized compressor units (MCUs) [8] placed at the clusters of producing wells. At this stage, the significantly increased electric load of well clusters fundamentally changes the nature of electric loads of gas fields, technical solutions for power supply sources and for the construction of the field's distribution grid.

On the example of the Vyngapurovsky field (shown in Fig. 1.) [9], [10] changes in actual gas withdrawals and dynamics of the calculated electrical load over time are considered.

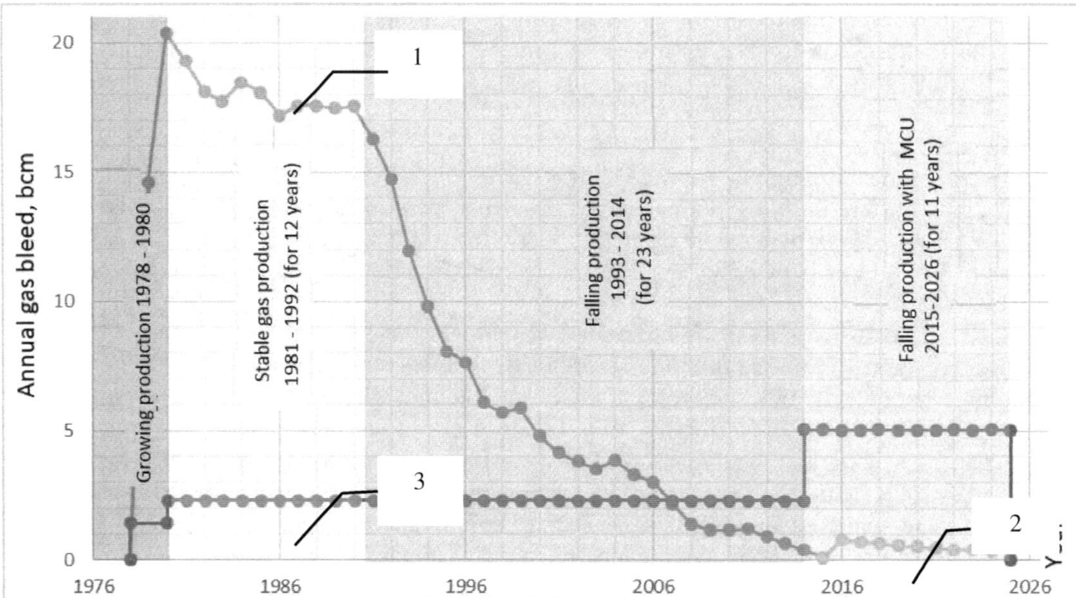

Figure 1. Annual gas sampling and dynamics of the calculated electric load at the Vyngapurovskoye gas field
1 is gas sampling curve before entering the MCU, **2** is gas sampling curve taking into account the MCU, **3** is estimated electric load of the field

The graph (Figure 1) shows that at the Vyngapurovsky field, the first period of "increasing production" lasted only 3 years: from 1978 to 1980, the second period of "constant production" – 12 years from 1981 to 1992, the third period of "falling production" – 34 years from 1993 to 2026. The third period is divided into two stages: before the MCU was commissioned (23 years - from 1993 to 2015) and after the MCU was commissioned (11 years-from 2015 to 2025).

Thus, the accumulated life cycle of the Vyngapurovsky gas field at the moment is:
- before the commissioning of the MCU-38 years;
- with the commissioning of the MCU-49 years.

Commissioning of the MCU allowed extending the life of the gas field by 11 years and increasing the accumulated gas extraction by 5.21 billion m3.

The graph shows an increase in the electric load during gas production in each period of the gas field life cycle, but the introduction of MCU radically changes the level of electric loads, their distribution over the field territory and, accordingly, requires a change in the distribution grid.

To implement the distributed compression technology from the MCU at the Vyngapurovsky field, it was necessary to reconstruct the existing power supply system in the following volume:
- reconstruction of existing overhead power lines (overhead lines) with a voltage of 6 kV and a length of 22 km;
- construction of a new closed switchgear with a voltage of 6 kV – 1 PC.;
- construction of a new point of automatic voltage regulation (PAVR) 6 kV - 1 PC.

The situation is much worse at the Yamburg oil and gas condensate field (OGCF). To implement the distributed gas compression technology, MCUs are being introduced at the field [11], [12], [13] in quantity-52 PCs. with a single power of 1 MVA.

To implement this technology at the Yamburg OGCF, it is necessary to reconstruct the entire existing power supply system in the following volume:
- construction of a 10 kV overhead line with a length of 410 km;
- construction of 35 kV overhead lines with a length of 124...169.2 km;

- construction of substation (SS) 10/35 kV-2 PCs.;
- construction of substation SS 35/10 kV-4...5 PCs.;
- reconstruction of existing 110 kV substations of the integrated unit of gas preparation plant No. 2 (UGPP2), UGPP 4, UGPP 6 - construction of a 35 kV CTTS.

The design of the power supply system for the Vyngapurovsky gas field and the Yamburg OGCF, was performed without taking into account electrical loads at the final stage of the field's life cycle.

Thus, one of the main tasks of the electric power engineering of the gas industry is to create an optimal power supply system at the design stage, taking into account the prospective growth of electric loads.

The task of this study is to evaluate the calculated electrical loads in each life cycle period for the existing gas fields in Western Siberia and determine their range of changes.

3. Theory

The external power supply system for gas fields consists of a supply and distribution grid.

Based on literature sources [5-7], [14-15] and the experience in designing of power supply systems for gas fields the paper calculates the electrical load and assesses its dynamics for the technological stages of gas field development.

In this study, an integrated approach was used to calculate the electrical load [16], which is performed based on the total annual power consumption W_{pc} and the average annual demand coefficient K_d.

The calculation of electrical loads is based on the indicators of the existing gas fields in Western Siberia.

During the entire life cycle of a gas field, its technological scheme changes.

Gas production volumes, average reservoir pressure, and bottom-hole pressures decrease over time from the first to the third period.

In order to maintain gas production volumes at the fields, the following is introduced: in the second period – BCS, in the third period – increasing the capacity of DCS and installing MCS at the mouth of gas wells [8].

The transformation of the technological scheme leads to a change in the composition of technological equipment in the gas field whose electric receivers are connected to the power supply system.

Three technological schemes of a gas field corresponding to the technological stages of its life cycle:

- "layer - borehole- GCCS - CGPP -main gas pipeline";
- "layer-borehole-GCCS- BCS - CGPP - main gas pipeline";
- "layer-borehole-MCU – GCCS – BCS - CGPP - main gas pipeline".

Each period of the life cycle of a gas field has its own consumption of electric energy.

All gas fields are characterized by continuous operation. Thus, knowing the actual electricity consumption (measurement data) and the annual number of hours of maximum load usage, it is possible to calculate the actual calculated electrical load using the expression:

$$P_{ac} = \frac{W_{ec}}{T_M}, \text{MV} \tag{1}$$

where W_{ec} is the actual electricity consumption per year, thousand kV h; T_M=7650 h is the annual number of hours of maximum use [17].

Table 1 shows the values of the actual annual electricity consumption and the actual electrical design load for 2019 for a number of gas fields.

Table 1. Annual electricity consumption and actual estimated load for 2019.

Period of life	Name of the field	Electricity consumption, thousand kVh per year	Actual design load, MV
Falling gas production	Medvezye	41 950.58	5.48
Falling gas production	Urengoy	154 494.80	20.20
Falling gas production	Yamburg	146 237.10	19.12
Falling gas production	Vyngapurovsky	40 850.96	5.34
Falling gas production	Komsomolsky	23 936.08	3.13
Falling gas production	Zapadno-Tarkosalinsky	20 233.40	2.64
Falling gas production	Gubkinsky	21 000.00	2.75
Falling gas production	Yubileyny	32 460.48	4.24
Falling gas production	Yamsoveyskoe	34 228.88	4.47
Falling gas production	Yuzno-Russky	38 209.93	4.99
Falling gas production	Vyngayakhinsky *	23 763.00	3.11
Falling gas production	Yety-Purovsky *		

* - one power source for two fields

According to the scheme of power supply of each field is estimated installed capacity of electrical systems 6(10) kV in each period of the life cycle of gas fields (table 2.). For electrical receivers of the first category of reliability of power supply is taken into account 50% load capacity of power transformers [18], for the third category - 100% load power. With a known installed capacity and an actual average annual design load, the average annual demand coefficient is determined:

$$K_d = \frac{P_{ac}}{P_e}, \text{ o.e.} \tag{2}$$

where Pac is the actual average annual calculated electric load, kV; Pe is the total established capacity according to the power supply scheme, kV. The average annual demand coefficient was calculated for each field, and the calculation results are shown in Table 2.

The resulting average annual demand coefficient is used to calculate the load during the remaining periods of the field life cycles

Table 2. The results of calculating the average annual demand coefficient.

No	The name of the field	Total installed capacity, kVA			Rated load, kV	Average annual demand coefficient, a.u.
		The first	The second	The third		
1	Medvezye	15 468.0	29 878.0	29 878.0	5 483.73	0.18
2	Urengoy *	-	-	-	20 195.40	-
3	Yamburg *	-	-	-	19 115.96	-
4	Vyngapurovsky	3 185.0	5 185.0	11 372.5	5 339.99	0.47
5	Komsomolsky	5 502.5	14 462.5	14 462.5	3 128.90	0.22
6	Zapadno-Tarkosalinsky	6 132.5	8 732.5	11 230.5	2 644.89	0.24
7	Gubkinsky	4 833.0	8 333.0	8 333.0	2 745.10	0.33
8	Yubileyny	8 560.0	14 810.0	19 810.0	4 243.20	0.29

9	Yamsoveyskoe	7 028.0	12 288.0	12 288.0	4 474.36	0.36
10	Yuzno-Russky	9 545.0	12 542.0	14 805.0	4 994.76	0.34
11	Vyngayakhinsky + Yety-Purovsky **	6 112.0	8 112.0	8 112.0	3 106.27	0.38

* - the average annual demand coefficient was not calculated
** - one power source for two fields

Changes in the electrical load in the second and the third periods of the life cycle of deposits relative to the first period are determined. thus load growth coefficients are obtained (Table 3). There are also coefficients of electric load growth by stages of the life cycle. These indicators can be used to select the voltage class and the scheme of a supply grid of the fields.

Table 3. Calculated electrical load in each period and load growth factors.

No	The name of the field	The estimated electrical load during the life cycle of a field. MVA			Coefficient of increase in the load of the supply network relative to the first period. a.u.	
		The fist	The second	The third	In the second period	In the third period
1	Medvezye	2.84	5.48	5.48	1.93	-
2	Urengoy *	8.77	11.42	11.42	1.30	-
3	Yamburg *	13.18	18.05	49.25	1.37	3.74
4	Vyngapurovsky	1.50	2.43	5.34	1.63	3.57
5	Komsomolsky *	1.19	3.13	3.13	2.63	-
6	Zapadno-Tarkosalinsky	1.44	2.06	2.64	1.42	1.83
7	Gubkinsky	1.59	2.75	2.75	1.72	-
8	Yubileyny	2.45	4.24	5.68	1.73	2.31
9	Yamsoveyskoe	2.56	4.47	4.47	1.75	-
10	Yuzno-Russky	3.22	4.23	4.99	1.31	1.55
11	Vyngayakhinsky + Yety-Purovsky	2.34	3.11	3.11	1.33	-

* - The estimated electric load of the field during the life cycle is determined by measurements of the annual electricity consumption separately for the CGPP and DCS
** - The MCU is assumed to be put into operation at Komsomolsky (21 units) [15]. It is not taken into account in calculation.

The found coefficients of electric load increase are applicable for choosing the voltage class of the field supply grid.

The voltage class of distribution grids of existing gas fields is 6. 10 kV. At these fields. gas well clusters were not initially electrified; the main consumers of electricity in the distribution grids of 6 and 10 kV of the gas collection and transport system were stations for electrochemical protection of gas pipelines. Electrification of gas well clusters began in 2000 with the introduction of unpopulated technologies and safety systems. i.e. the main consumers of electricity on the clusters were automated process control devices.

The transformer capacity of gas well clusters in the first and second periods of the field's life. as well as in the first stage of the third period. is from 25 to 100 kV.

According to the current Standart of Organization of Gazprom 2-6. 2-1028-2015 [18] gas well clusters belong to the third category of power supply reliability.

Electric loads of a field distribution grid during the third period of the life cycle before the introduction of compressor production technology practically do not change. The load growth rate with the introduction of the MCU. i.e. at the second stage of the third life cycle period. is determined for two fields where the MCU is installed or is expected to be installed - Vyngapurovsky and Yubileyny. The calculation of the average annual demand coefficient for a distribution grid of these fields is presented in Table 4.

Table 4. Average annual demand coefficient for a distribution grid

No	Distribution grid object	The total installed capacity in stages. kVA			Actual calculated electrical load. kV	Average annual demand ratio a.u.
		The first	The second	The third		
		Vyngayakhinsky				
1	OHL-6 кV "EPS Yug 1"	90	90	1 890	1 197.00	0.61
2	OHL-6 кV "EPS Yug"	128	128	1 028	852.40	0.62
3	OHL-6 кV "EPS Severo-Zapad"	266	266	1 806	1 675.30	0.62
4	OHL-6 кV " Severo-Vostok"	151	151	601	495.80	0.64
		Yubileyny				
5	V34-Cluster 33	510	510	3 010	13 035.60	0.57
6	V35-Cluster 25	450	450	2 950	12 852.00	0.57

The electric power of the MCU is from 450 to 1000 kW. The electrical loads of the distribution grid and load growth factors in the second and third periods of the field life cycle are presented in Table 5.

Table 5. Estimated electrical load of the distribution grid and load growth factors at the second stage of the third life cycle period.

No	The name of the field	Estimated electrical load during the life of the field. kV			Coefficient of increase in the load of the distribution grid relative to the first period. a.u.	
		The first	The second	The third	Во втором периоде	В третьем периоде
		Vyngayakhinsky				
1	OHL-6 кV "EPS Yug 1"	54.82	54.82	1197.0	1	21.83
2	OHL-6 кV "EPS Yug	7918	7918	852.4	1	10.77
3	OHL-6 кV "EPS Severo-Zapad"	164.83	164.83	1675.3	1	10.16
4	OHL-6 кV " Severo-Vostok"	96.48	96.48	495.8	1	5.14
		Yubileyny				
5	V34-Cluster 33	256.27	256.27	1680.0	1	6.56
6	V35-Cluster 25	288.72	288.72	1704.0	1	5.9

Table 5 shows a multiple (from 5.14 to 21.83 times) increase in the load of the distribution grid at the second stage of the third period of the gas field life cycle.

4. Analysis of results

Analysis of changes in the calculated electrical load in each period of the life cycle at the existing fields in Western Siberia and comparison of loads relative to the first period allowed us to establish the coefficient of increase in the calculated average annual electrical load capacity, which is as follows:

- for the supply line in the second period of life, it is from 1.37 to 2.63 a.u. (due to the introduction of BCS);
- for the supply line in the third period, it is from 3.03 to 3.74 a.u. (due to the input of BCS and MCU);
- for the distribution grid for gas well clusters at the second stage of the third period, it is from 5.9 to 21.83 a.u. (due to the introduction of MCU).

5. Discussion of results

The rapid electrical load growth leads to the need for a radical reconstruction of the distribution grid of deposits at the final stage of its life cycle. Such reconstruction will require not only the reconstruction and additional construction of 6(10) kV transmission lines, but also the transfer of the field distribution grid to a higher voltage class (20 and / or 35 kV)

The growth of the electric load with a decrease in reservoir pressure during the life cycle of a gas field obtained by estimating the dynamics of the electric load is shown in Fig. 2.

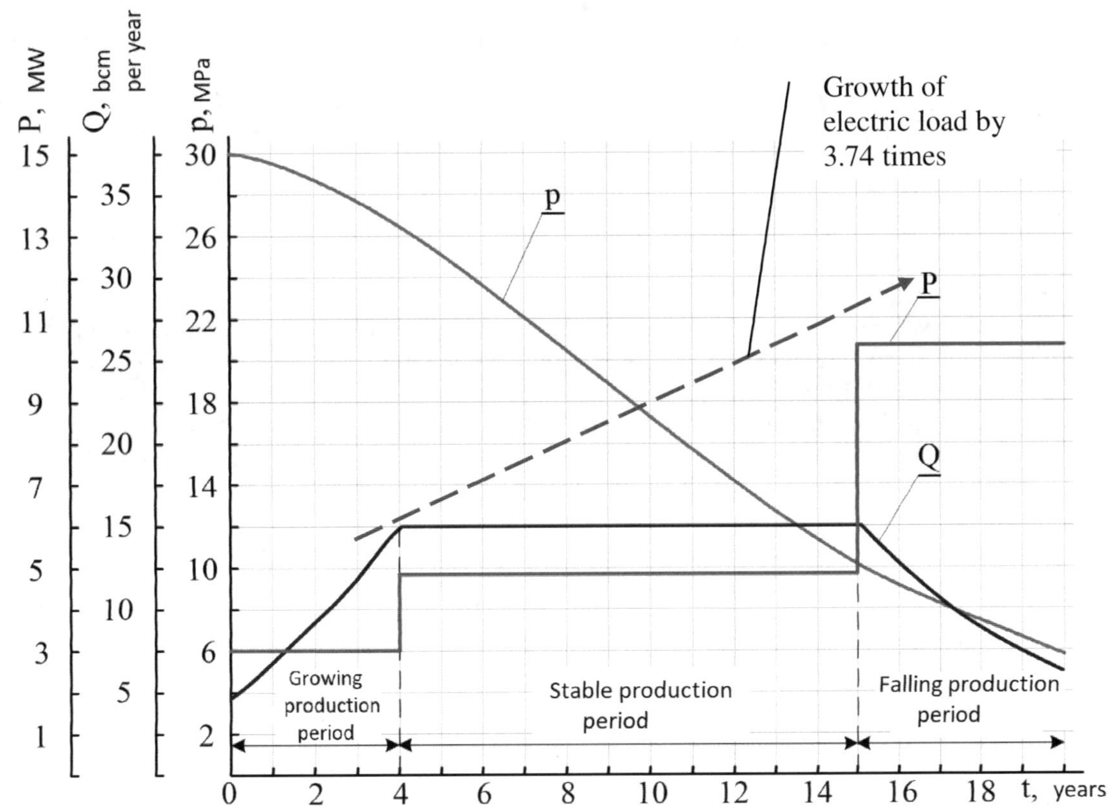

Figure 2. Technological indicators of field development and estimated electric capacity.
Q is annual gas production; p is reservoir pressure; P is average annual estimated electrical load capacity.

6. Conclusion

During the life cycle of a gas field, with a decrease in well flow rate and average reservoir pressure with the introduction of compressor technology for gas production, the calculated electrical load increases many times.

The introduction of gas production technology from MCU requires a dramatic reconstruction of the power distribution network of gas fields with a transition to a higher voltage class – 20 and 35 kV. Reconstruction of the power supply grid and development of power sources may also be required.

A multiple increase in electricity consumption leads to a significant increase in the specific share of energy consumption in the cost of gas production at the final stage of the life cycle of Western Siberia fields. With this in mind, all solutions for the reconstruction of the gas field power supply system with the introduction of MCU technology require detailed and qualified feasibility studies.

References

[1] Kudryashov R A 2009 Improvement of calculations of power consumption and electric loads in the changed conditions of designing power supply for oil and gas fields *J. Oil industry* **2** 56-58

[2] Kudryashov R A, Novoselov Y B and Frajshteter V P 2003 The Methods for calculating specific power consumption in oil production *J. Energy of the Tyumen region* **1** 10-15

[3] Kudryashov R A, Novoselov Y B, Frajshteter V P and Evseenko D V 2002 The Justification of power consumption levels and electrical loads in the design of power supply for oil fields *J. Oil industry* **7** 39-41

[4] Kudryashov R A and Kudryashova O M 2004 The role of the methodological factor in refining the estimated maximum capacity for oil-producing enterprises *J. Oil industry* **3** 82-83

[5] Kudryashov R A, Malkova Z A, Novoselov J B and Frajshteter V P 2004 Regulatory framework for designing power supply for oil fields *J. Oil industry* **3** 76-78

[6] Kudryashov R A 2014 Results of research on power consumption and electrical loads at oil and gas industry facilities in Western Siberia *J. Oil industry* **2** 52-55

[7] Kudryashov R A, Novoselov Y B and Frajshteter V P 1986 Reasons for overestimation of calculated electrical loads in the oil fields of Western Siberia *J. Industrial power engineering* Issue **6** 18-21

[8] Vorontsov M A, Rotov A A, Marushchenko I V and Laptev E M 2014 The Prospects for the use of straightened compression in field gas production systems *J. Vesti gazovoi nauki: nauch. teh.digest* **4 (20)** 164-73

[9] Saranca A B 2014 Low-pressure gas from the Cenomanian deposits of the Yamal-Nenets Autonomous district *Academic J. of Western Siberia* **10 3(52)** 146-147

[10] Saranca A B and Ouay V A 2016 Application of Russian-made modular compressor units for low-pressure Cenomanian gas production *Academic J. of Western Siberia* **10 1(62)** 21 - 22

[11] Minlikaev V Z, Dikamov D V, Arno O V, Merkulov A V, Kirsanov S A, Krasovsky AV, Svensky S Yu and Kononov A V 2015 Application of mobile compressor units at the final stage of development of gas *J. Gas Industry* **1(717)** 15-17

[12] Arno O B 2015 Technical and technological solutions and innovations at different stages of the life cycle of Yamburg deposits *Sc. J. of Russian newspaper society* **2-3** 7-14

[13] Serotetto I V and Navigatin V V 2015 Equipment and technology for construction and repair of oil and gas wells / Machinery and equipment for oil and gas fields (Tyumen: Industrial University of Tyumen) pp 169-72

[14] RTM 34.18.32.4-92 1993 Instructions for calculating electrical loads. Instructive and informational materials on the design of electrical installations (Moscow: "Tyazhpromelektroproekt")

[15] Semushkin AV and Mikheev S S 2015 Problems and prospects of development of the Komsomolsk and West Tarkosalinsky GP booster complex *Gas transmission systems: present and future Gas transmission systems: present and future: Abstracts of the VI International scientific and technical conference GTS-2015* p 47
[16] Sazykin V G and Kudryakov A G 2017 Calculation of electrical loads in the power supply systems of the agro-industrial complex: method. instructions (Krasnodar: Kuban state agrarian University) p 54
[17] Standards of technological design EPP-94 Design of power supply for industrial enterprises 1994 (Moscow: "Tyazhpromelektroproekt") p 25
[18] Standart of Organization of Gazprom 2-6.2-1028-2015 Categorization of electric receivers of industrial facilities of LLC "Gazprom" 2015 (Moscow: without publishing house) p 65

Studying Interaction of Multi-Motor Forming System for Double Curvature Items

N E Deryuzhkova, V A Solovyev and V V Teterin
Electrical Drive and Industrial Unit Automation Department, Komsomolsk-na-Amure State Technical University, Lenin Avenue, 27, Komsomolsk-na-Amure, 681013 Russia

E-mail: epapu@knastu.ru

Abstract. Large double curvature panels are widely used in manufacture of modern aircraft bodies. One of the possible ways to manufacture such parts is to form a workpiece using the method of high temperature multi-point forming by means of a set of moving cores. The paper considers workpiece forming process using different core unit configurations and presents the results of workpiece behaviour study during forming, obtained by means of a finite element method. Conclusions are made on the existing interconnections between forming cores during the forming process.

1. Double Curvature Parts Manufacture Methods

Large load-bearing structural members, forming complex aero- and hydrodynamic circuits of air- and watercrafts, are usually very thick parts made of extra-strong materials (aluminium, steel, titanium) and having a complex spatial form. Mechanical treatment methods are widely used in their manufacture when a part of required shape is milled from large workpieces (forgings). This process is relatively simple, it can have many transitions, including heat treatment, a low material utilization rate, a large amount of material goes into chips, and as a result, the process is long and expensive. Another method of manufacturing large parts of complex spatial form is a deformation process. However, ensuring high forming accuracy of large and thick double curvature parts using traditional methods is a difficult and, sometimes, even impossible task. Existing solutions for forming large plates from high-strength alloys based on multiple step forming and/or multiple die punching have their limitations on press equipment, dimensions and loads as well as high cost of dies. One of the most promising solutions is deforming workpieces using multi-point forming by means of a reconfigurable core punch (matrix).

2. Existing Solutions

Reconfigurable core equipment predominantly has fixed matrix and punch forms. Constant and equal speeds of cores make it impossible to ensure optimal deforming speed, which means that double curvature parts need to be formed in several steps.

There is another approach to reconfigurable systems based not on the geometrical versatility of equipment but on the optimization of deformation modes, which includes deforming speed control and process intensification by heating the part. Many technical solutions implementing this approach [1-9] try to preserve physical and mechanical properties of formed material, minimize residual stress and avoid ductile fractures.

Content from this work may be used under the terms of the Creative Commons Attribution 3.0 licence. Any further distribution of this work must maintain attribution to the author(s) and the title of the work, journal citation and DOI.
Published under licence by IOP Publishing Ltd

For this, forming is performed in creep mode where it is possible to maintain constant forming speed at each point. Various technical devices are suggested to provide required forming modes: individual core electric drives, differential motors based on hydraulic cylinders, load equalizing devices, etc.

Core unit composition and structure are determined by a set of process steps required to form large plates, their sequence and peculiarities.

The unit consists of three subsystems:

Workpiece treatment subsystem. It includes main functional parts, which actually perform process steps on a workpiece (preheating chamber, forming chamber, prequenching heating chamber, quenching chamber).

Support subsystem. It contains auxiliary functional parts, which ensure uninterrupted operation of main parts.

Monitoring and control subsystem. It contains hardware and software controlling all workpiece treatment and support processes.

Forming cores are two sets of opposing blocks. Core moving drive is based on autonomous variable-frequency electric drives. Rotational to linear motion conversion is performed by actuator that is an electric cylinder with a standard ball screw. Reduction gearbox is used to coordinate operation of electric motor and actuator.

The forming chamber is a system of two prefabricated sections, partially welded and partially assembled using bolt joints. Each section consists of two bearing plates: upper and lower. Upper and lower bearing plates are connected by the main basic frame members. Blocks are secured to the upper and lower bearing plates with bolt joints.

The main task is studying workpiece behaviour during multi-point forming to determine the optimal mode of forming core drive which ensures high item quality.

3. Forming Process Study

First, let us take three-point bend as a base part to study forming process. Three-point bend process is schematically shown in figure 1.

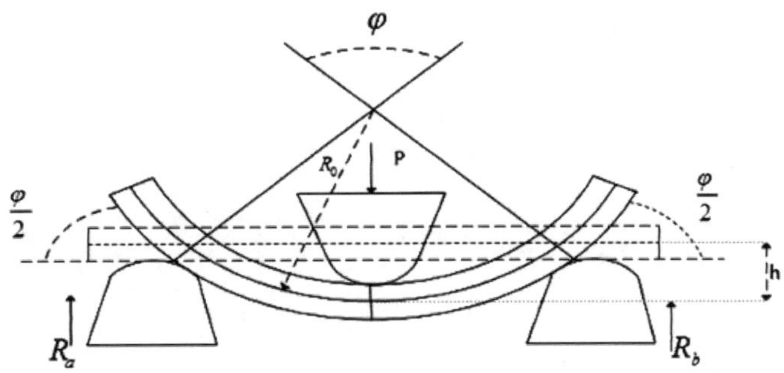

Figure 1. Bending Process Illustration

Designations in figure 1 correspond to the following: R_0 is assumed central radius; φ is bend central angle; h is deflexion value (linear forming cores movement); P is force affecting a workpiece; Ra, Rb are reactions at the opposing cores.

Let us determine the forming force with the following expression [10], H:

$$\Delta P = \frac{\sigma_s \frac{Bs^2}{4} + \frac{\Pi}{R_0}\frac{Bs^3}{12}}{\frac{l}{4} + \frac{\Delta h}{4}\frac{tg(\frac{\varphi}{2}) - \mu}{1 + \mu tg(\frac{\varphi}{2})}}$$

where s is workpiece thickness, m; B is workpiece width, m; R_0 is bend central radius, m; l is distance between opposing cores axes, m; σ_S is creep limit, Pa; Π is material hardening modulus, Pa; μ is friction factor between core and workpiece.

Bend angle φ changes during workpiece forming from zero to some final value determined by linear movement. At a first approximation, tg φ/2 can be expressed through forming part parameters tg φ/2 = 2Δh/l [11].

$$\Delta P = \frac{\sigma_S \dfrac{B \cdot s^2}{4}}{\dfrac{l}{4} + \dfrac{\Delta h}{2} \cdot \dfrac{2 \cdot (\Delta h / l) - \mu}{1 + 2 \cdot \mu \cdot \Delta h / l}}.$$

Forming force can also be decreased by heating the workpiece which allows lowering creep limit and reducing hardening to zero. It should be noted that bending accuracy increases in this case due to reduced springback effect.

Now let us determine workpiece behaviour during forming. Finite element method is used for this. Workpiece bending process is simulated in Femap [12]. Let us consider two cases: during operation of the central core (figure 2) and one of the side ones (figure 3). It should be noted that when one core moves, others are still.

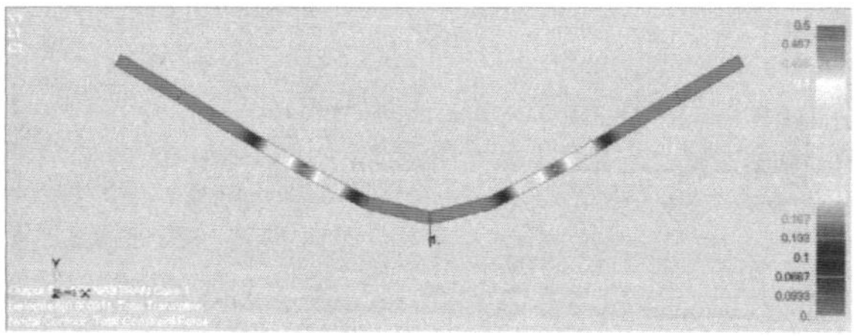

Figure 2. Workpiece behavior during central during core operation

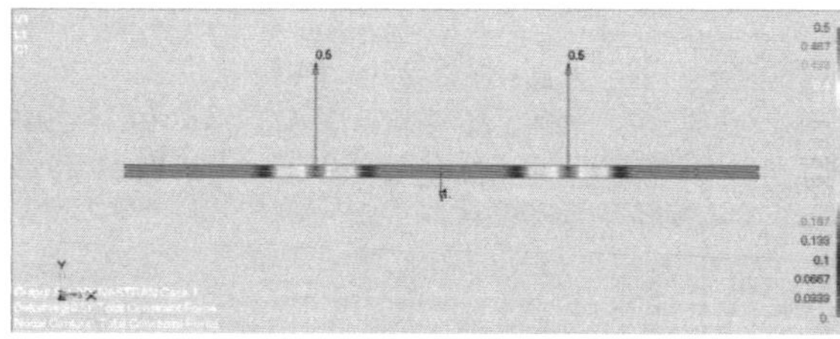

Figure 3. Forming force and support reactions central core operation

Figure 3 shows the values of non-moving cores support reactions when central core moves. Meaning, that if central core presses down on the workpiece with force P, then the workpiece applies force Q = P/2 to each of the side cores.

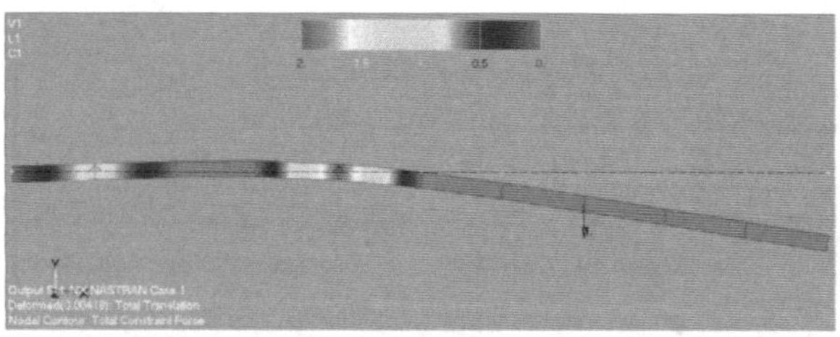

Figure 4. Workpiece behaviour during operation of one of the side cores

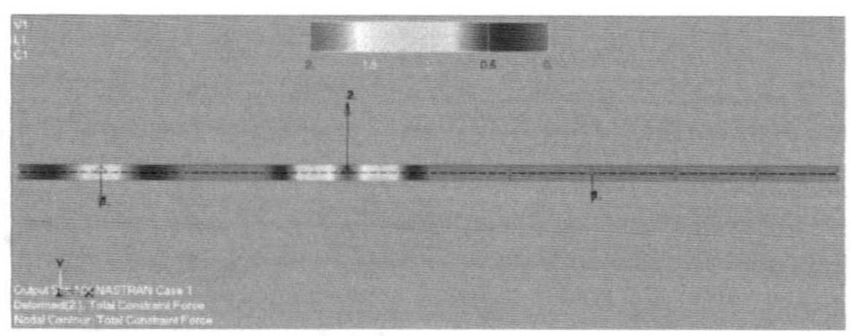

Figure 5. Forming force and support reactions during operation of one of the side cores

Figure 5 shows the values of non-moving cores support reactions when one of the side cores moves. Meaning, that if a side core presses down on the workpiece with force P, then the workpiece applies force Q1 = 2*P to the central core and Q2 = P to another (non-operating) side core.

When cores move simultaneously, each of them is affected by adjacent cores movement and affects them in turn. Central core takes most of the load in this bending scheme. If cores operate separately, there is no significant interaction.

The effect of deformation speed on forming process should also be taken into account. If the selected deformation speed is too high, there may be indentations on the workpiece from forming cores, ductile fractures due to stress in the workpiece which exceed yield limit [13]. If deformation speed is too low,

IOP Conf. Series: Materials Science and Engineering 1118 (2021) 012016 doi:10.1088/1757-899X/1118/1/012016

it unjustifiably lengthens the process. Too high deformation speed also increases the load on the forming rod.

When hot-forming method is used, two processes occur simultaneously and affect deformation resistance in opposite directions: hardening (cold work) and softening (recrystallization). Both processes take place in time. Hardening speed is determined by deformation speed and softening speed — by recrystallization speed, which depends on metal temperature. Depending on the ratio of deformation and recrystallization speeds (hardening and softening), deformation resistance at this temperature changes: the faster the deformation speed, the higher the deformation resistance. Deformation resistance at this speed and temperature is affected by deformation degree: the higher the deformation degree, the greater the cold work and deformation resistance [14, 15].

Therefore, deformation speed and degree directly increase deformation resistance and indirectly decrease it. When hot-forming method is used and heat output is small, hardening effect of both factors prevails, and deformation resistance rises with the increase of deformation speed and degree [16, 17].

Differentiation of thermal power load modes by deformation speed allowed to empirically establish that deforming at deformation speeds with order $10^{-3}...10^{-4}$ s^{-1} has certain advantages compared to fast ("instant") elastoplastic deformation: forming force is reduced significantly, material deformation index increases. Moreover, during bending and relaxation calculations, area of elastoplastic stresses has to be monitored for active or unloading processes at the points of workpiece elements, as irreversible deformations of plasticity have to be taken into account, which significantly complicates the calculations [18, 19].

In view of the above, control object model in Simulink can be presented as shown in figure 6 [20].

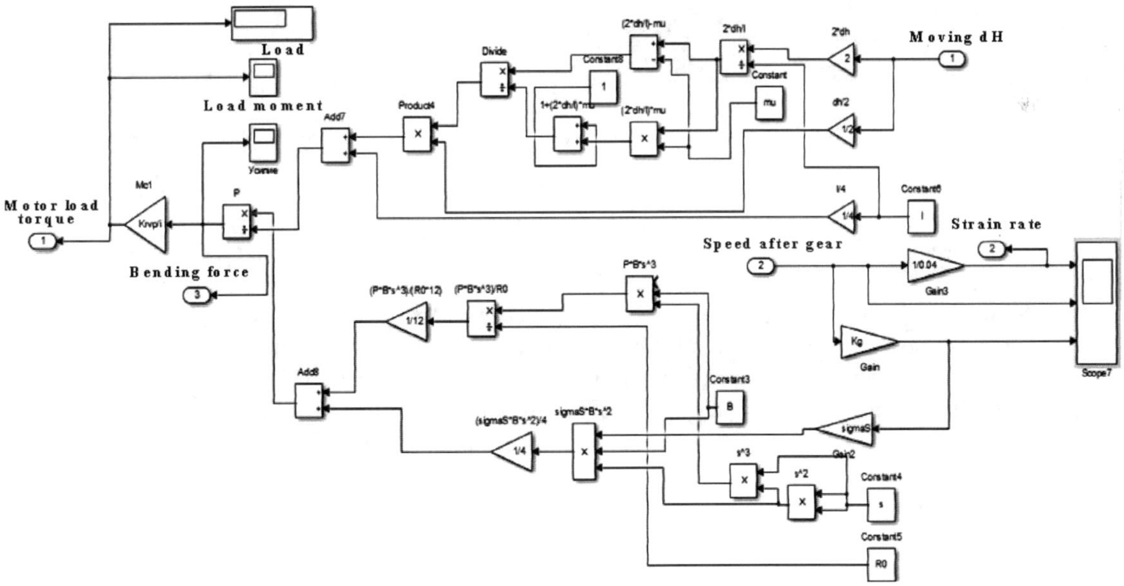

Figure 6. Control Object Model

Three-point bending part is basic. Most important is the possibility of forming sheet materials. Reconfigurable forming core matrix may be used for this. Let us study workpiece behaviour at different quantities of forming cores.

Figure 7 shows force distribution by core with matrix 3x3 layout. Central core presses down with P = 1 force, and outer cores experience 0.384*P reaction.

Figure 8 shows Matrix 5x5 where adjacent cores experience 0.35*P reaction during central core operation.

Figure 7. Matrix 3x3

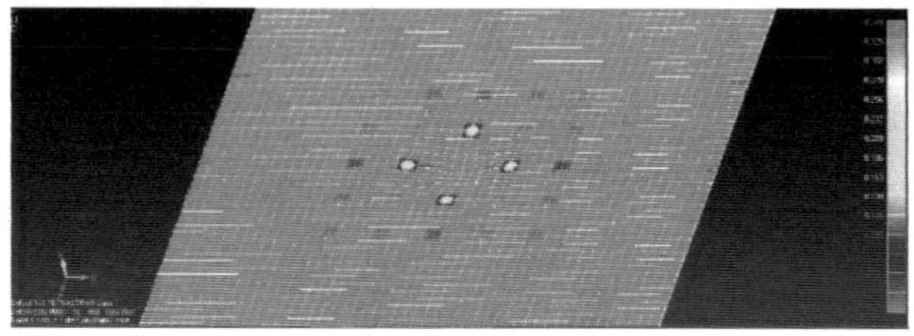

Figure 8. Matrix 5x5

Figure 9 shows Matrix 11x11 where adjacent cores experience 0.351*P reaction during central core operation.

Further increase of matrix size does not lead to significant force change. Matrix 5x5 is selected for study as it has sufficient accuracy and processing obtained results will not take a lot of time.

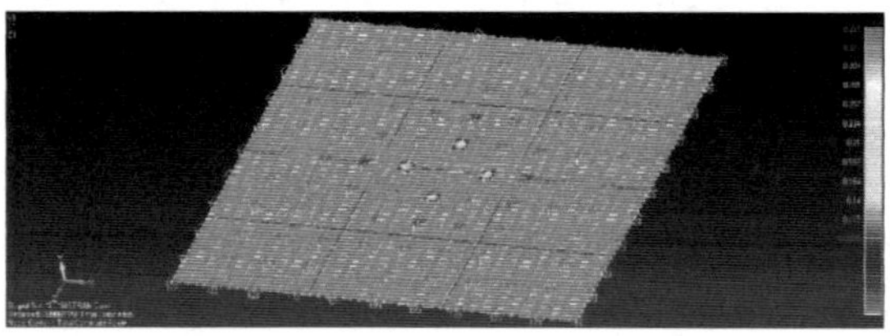

Figure 9. Matrix 11x11

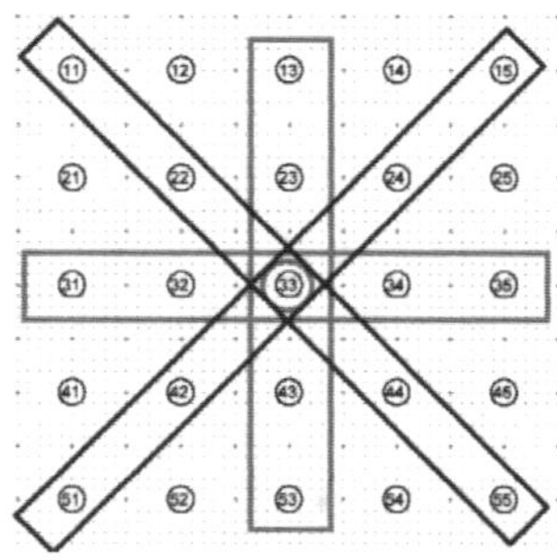

Figure 10. Force Distribution in Matrix 5x5

During the study, it was established that during operation of one of the cores, the highest load falls on adjacent cores; in particular, ones located at matrix row and column lines (figure 10). Cores located in red area experience the highest load during core 33 operation, cores in blue area — the lowest. Other cores are in intermediate positions.

Now let us determine how movement of one core affects other adjacent cores. For this, we simulate each core operation using finite element method and estimate its impact on other cores. It should be noted, that operating core moves during simulation, others stay in place.

As the result of simulation, special points under positions 11 and 33 were determined, i.e. outer and central. Let us study matrix core movement in these cases as they are extreme and other cases are transient.

What distinguishes the outer point is that maximum force impacting adjacent cores is 0.94*P which is similar to bending force. Maximum force of operating core under position 33 is 0.35*P which is almost one third of bending force.

As in the case of three-point bending, core interaction and, consequently, drive interaction manifests only for simultaneous core operation. Meaning, that load moment for core drive is composed not only from the required bending force (moment) but from load moments occurring due to adjacent core drive operation (support reaction).

In general, if each core drive develops forming force,

$$P = \begin{pmatrix} P_{11} & P_{12} & P_{13} & P_{14} & P_{15} \\ P_{21} & P_{22} & P_{23} & P_{24} & P_{25} \\ P_{31} & P_{32} & P_{33} & P_{34} & P_{35} \\ P_{41} & P_{42} & P_{43} & P_{44} & P_{45} \\ P_{51} & P_{52} & P_{53} & P_{54} & P_{55} \end{pmatrix},$$

then for each core force is as follows: $$P_{ROD} = \sum_{i=1}^{5} \left(P_{1,i} + P_{2,i} + P_{3,i} + P_{4,i} + P_{5,i}\right).$$

For example, for core 33 (central), load force is determined as follows:

$$\begin{gathered} P_{ROD33} = P_{33} + 0.35 \cdot P_{32} + 0.35 \cdot P_{34} + 0.35 \cdot P_{23} + 0.35 \cdot P_{43} + \\ 0.2 \cdot P_{53} + 0.2 \cdot P_{13} + 0.2 \cdot P_{31} + 0.2 \cdot P_{35} + \\ 0.03 \cdot P_{15} + 0.03 \cdot P_{51} + 0.03 \cdot P_{55} + 0.03 \cdot P_{11} + \\ 0.008 \cdot P_{24} + 0.008 \cdot P_{42} + 0.008 \cdot P_{44} + 0.008 \cdot P_{22} + \\ 0.082 \cdot P_{21} + 0.082 \cdot P_{12} + 0.082 \cdot P_{14} + 0.082 \cdot P_{41} + \\ 0.082 \cdot P_{25} + 0.082 \cdot P_{52} + 0.082 \cdot P_{45} + 0.082 \cdot P_{54}. \end{gathered}$$

Here force P_{33} is forming force during separate core operation, other constituents manifest only when all cores move together. These expressions can be obtained for each core. All cores will not necessarily move. If one core is still, its force P_{ij} is taken as zero.

In view of the above, a conclusion can be made that in case of simultaneous movement of all cores, their drives need to be ready to process higher load than in case of separate operation. Simultaneous movement of all cores allows giving workpiece the required form right away instead of doing it in stages (as in case of separate operation). The results obtained using finite element method are approximate but adequate.

If we suppose that each core develops the same forming force P, then each core will experience the following force during their simultaneous operation:

$$P_{ij} = \begin{pmatrix} P_{11} & P_{12} & P_{13} & P_{14} & P_{15} \\ P_{21} & P_{22} & P_{23} & P_{24} & P_{25} \\ P_{31} & P_{32} & P_{33} & P_{34} & P_{35} \\ P_{41} & P_{42} & P_{43} & P_{44} & P_{45} \\ P_{51} & P_{52} & P_{53} & P_{54} & P_{55} \end{pmatrix} = P \cdot \begin{pmatrix} 1.943 & 3.161 & 3.028 & 3.161 & 1.944 \\ 3.161 & 4.252 & 4.233 & 4.255 & 3.159 \\ 3.028 & 4.233 & 4.059 & 4.232 & 3.031 \\ 3.161 & 4.253 & 4.233 & 2.252 & 3.16 \\ 1.943 & 3.161 & 3.028 & 3.161 & 1.943 \end{pmatrix},$$

This way, simultaneous operation of all cores leads to increased load on electric drive. Central core drives are most affected. This is why electric drive system should be designed, taking into account possible load increase during simultaneous operation of all cores.

4. Conclusion

Multi-point forming by means of a reconfigurable core punch (matrix) is the most promising existing solution to manufacturing large double curvature parts. This method provides high accuracy and quality of manufactured parts.

Finite element method was used to study forming process. It was established that in case of simultaneous (parallel) movement of each core, additional reaction impacts adjacent cores which manifests as increased load on core drive. However, in case of separate (sequential) movement of cores, additional load does not affect electric drives of adjacent cores.

In other words, interaction between cores by workpiece manifests only when they move simultaneously. This fact should be taken into account during core unit design.

References

[1] Sosnin O V, Shubin I A, Gorev B V, Raevskaya G A 1985 Method of Forming Double Curvature Parts and Device for its Implementation. Inventor's certificate 1147471 USSR, IPC7 B21D 11/20 (USSR), No. 3593076/25-27.

[2] Rublevskiy L L, Sosnin O V, Gorev B V, Bondarenko M I 1994 *Forming Device* Patent 2076010 Russian Federation, No. 94044707/08

[3] Gusev G A, Lyubashevskaya I V, Rublevskiy L L 2001 *Forming Device* Patent 2216421 Russian Federation, No. 2001111132/02

[4] Klopotov I D, Lyubashevskaya I V, Raevskaya G A, Rublevskiy L L, Sosnin O V 2002 *Forming Device* Patent 2251464 Russian Federation, No. 2002119982/02

[5] Miodushevskiy P V, Raevskaya G A, Sosnin O V 1996 *Method of Part Forming and Device for its Implementation* Patent 2056197 Russian Federation, No. 5037750/08

[6] Gusev G A, Lyubashevskaya I V, Rublevskiy L L 2001 *Forming Device* Patent 2216422 Russian Federation, No.2001114316/02

[7] Lyubashevskaya I V, Rublevskiy L L 2003 *Forming Device* Patent 2275267 Russian Federation, No. 2003123899/02

[8] Bratukhin A G, Sirotkin O S, Bogolyubov V S, Karpov A V 1999 *Multi-purpose Automated Unit for Forming Large Plate Parts (yay) and Method of Configuring its Surface* Patent 2133163 Russian Federation, No. 97121154/02

[9] Gorev B V 2008 Technology for Forming Large Parts from Sheets and Plates in the Creep Mode. *Mechanical-Engineering Technology* **2** 11-17

[10] Gorbunov M N 1970 *Methods of Workpiece and Punching Works in Aircraft Manufacture* (Moscow.: Mechanical Engineering)

[11] Deryuzhkova N, Solovyev V and Chzho Aung Htet 2017 *Problem of Developing a Mathematical Model for an Object of Interconnected Bulk Forming System* (Komsomolsk-on-Amur: Scholarly Notes of Komsomolsk-na-Amure State Technical University)

[12] Peresypkin K, Peresypkin V and Ivanova E 2012 *Space Rocket Structures Modeling Using Finite element Method in MSC.Nastran* (Digital textbook. Ministry of Science and Higher Education of Russia, Samara State Aerospace University named after S.P. Korolev)

[13] Romanovskiy V 1979 *Cold Punching Reference Book* (Mechanic Engineering: Leningrad)

[14] Zubov V, Stepanov G and Shirokov A 2003 Deformation Speed Effect on Creep Limit of Steels of Various Strength. *Strength Problems* **5** 113-121

[15] Stepanov G 1991 *Elastoplastic Deformation and Destruction of Materials at Pulse Loading* (Naukova Dumka/Kiev)

[16] Ogorodnikov V, Tyunki E and Ivanov A 1995 Strength and Viscosity of Metals in Wide Deformation Speed Range. *Applied Mechanics and Theoretical Physics* **3** 134-140

[17] Stepanov G, Zubov V, Olisov A and Tokarev V 2000 Strength of Thin Metal Sheets at Tensile Impact. *Strength Problems* **4** 62-69

[18] Illarionov A and Popov A 2014 *Technological and Operational Characteristics of Titanium Alloys* (Ekaterinburg: Ural University Printing House)

[19] Vasilchenko S A et al. 2016 *Development of Technology and Equipment for Manufacturing Large Seamless Ribbed Panels*: Explanatory Note (Komsomolsk-na-Amur: Federal State Budgetary Institution of Higher Education Komsomolsk-na-Amure State University)

[20] Diakonov V 2002 *Simulink 4. Specialized Reference Book* (Saint Petersburg: Piter)

IOP Conf. Series: Materials Science and Engineering 1118 (2021) 012017 doi:10.1088/1757-899X/1118/1/012017

Comparison of control methods for inverted 2-degree of freedom pendulum mounted on the cart

O Y Sumenkov and A S Belyaev
Tomsk Polytechnic University, 30, Lenin Avenue, Tomsk 634050, Russia

E-mail: oys5@tpu.ru

Abstract. The paper considers and analyzes the existing methods of controlling systems with a deficit of control actions. Based on the analysis, it was decided to use linear controllers with feedback on the state vector of the system and to compare these methods - modal and linear-quadratic. For the study, a dynamic model was chosen, which is an inverted 2-degree of freedom (2DOF) pendulum mounted on a cart. By an iterative method, acceptable system of generalized coordinates that definitely describes the state of the model was selected. Kinematic relations that describe the position of the model in generalized coordinates were derived. Using the Lagrange procedure, a system of nonlinear differential equations describing the motion of a dynamic model was obtained. The procedure of model linearization about the upright equilibrium point was also carried out in order to synthesize control system. Based on the results of modeling, which was carried out by numerical integration method in the Matlab environment, conclusions were drawn on the applicability of these control methods and their effectiveness.

1. Introduction

The stabilization of the position of an inverted pendulum on a cart is one of the most common problems in control theory, since it is an unstable nonlinear system which has some of the degrees of freedom that are not directly controlled. Such complexities of the system lead to the fact that simple controllers [1-3], for example, PID-controller, do not provide the necessary transfer characteristics, since minimal dynamic error and settling time are required to stabilize the position of the pendulum.

The most common methods used to solve this problem are: modal regulator, linear-quadratic regulator [4-6], regulators based on artificial intelligence methods: genetic algorithms [7-9], fuzzy logic [10-14], neural networks [15-18]. However, neural network regulator needs a large amount of training data and there is an uncertainty of the choosing of its architecture. To implement genetic algorithms, time-consuming calculations are required. And fuzzy logic, although it is often applicable, is difficult to analyze for stability, and the synthesis of rules is usually a difficult task. To adjust the modal and linear-quadratic regulators, a mathematical model of the control object is required, according to which the roots of the stabilization system are adjusted in such a way as to provide the preassigned values of settling time and dynamic error. However, the advantage of the linear-quadratic controller is optimality because of its attachment to optimal control methods, which allows it to achieve better performance than the modal controller [6].

The purpose of this research is to compare the application of modal and linear-quadratic controller in the problem of stabilizing the upright position of an inverted 2DOF pendulum on a cart.

Content from this work may be used under the terms of the Creative Commons Attribution 3.0 licence. Any further distribution of this work must maintain attribution to the author(s) and the title of the work, journal citation and DOI.
Published under licence by IOP Publishing Ltd

2. The dynamic model of inverted 2DOF pendulum

To conduct a research, the dynamic model was chosen as it shown below in Fig. 1. In this model inverted 2DOF pendulum on cart was chosen with several assumptions: there is no movement of the trolley along Z axis, there is no rotation of the pendulum around its axis, cart and pendulum are absolutely solid bodies, and center of gravity coincides with geometric center of the cart. Then, to describe the laws of its motion, we use 4 generalized coordinates: θ, φ, x, y.

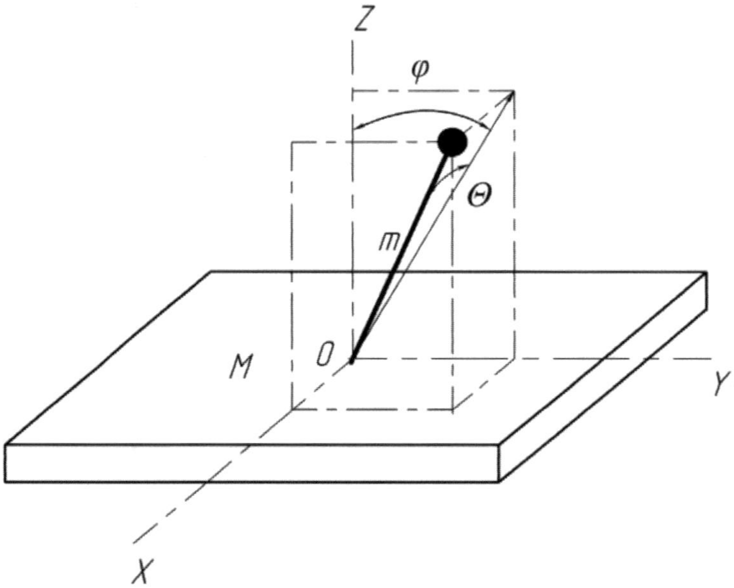

Figure 1. The dynamic model of an inverted 2DOF pendulum mounted on a cart

Here θ – is the angle of deflection of the pendulum from the ZOY plane, φ – is the angle of rotation of the pendulum in the ZOY plane, the zero position is measured from the Z axis, M – is the mass of the cart, m, l – is the mass and length of the pendulum, respectively, the force of gravity g is directed in the opposite direction to the Z axis.

Then we describe the location of the center of gravity of the pendulum in terms of generalized coordinates:

$$\begin{cases} x_p = x + l\sin\theta, \\ y_p = y + l\cos\theta\sin\varphi, \\ z_p = 0 + l\cos\theta\cos\varphi. \end{cases} \Rightarrow \begin{cases} \dot{x}_p = \dot{x} + l\dot{\theta}\cos\theta, \\ \dot{y}_p = \dot{y} - l\dot{\theta}\sin\theta\sin\varphi + l\dot{\varphi}\cos\theta\cos\varphi, \\ \dot{z}_p = -l\dot{\theta}\sin\theta\cos\varphi - l\dot{\varphi}\cos\theta\sin\varphi. \end{cases} \quad (1)$$

2.1. Nonlinear equations of inverted 2DOF pendulum

To describe the motion of complex mechanical systems with holonomic constraints, the Lagrange equations of the second kind are usually used [19-20].

To obtain the equation of the dynamics of the system using the Lagrange equations of the second kind, it is necessary to find expression for the kinetic energy, which for this system takes the form:

$$T = T_p + T_c = \frac{1}{2}m(\dot{x}_p^2 + \dot{y}_p^2 + \dot{z}_p^2) + \frac{1}{2}M(\dot{x}^2 + \dot{y}^2) = \frac{1}{2}M(\dot{x}^2 + \dot{y}^2) +$$

$$+\frac{1}{2}m((\dot{x} + l\dot{\theta}\cos\theta)^2 + (\dot{y} - l\dot{\theta}\sin\theta\sin\varphi + l\dot{\varphi}\cos\theta\cos\varphi)^2 + (-l\dot{\theta}\sin\theta\cos\varphi - l\dot{\varphi}\cos\theta\sin\varphi)^2) =$$

$$=\frac{1}{2}m(\dot{x}^2+\dot{y}^2+l^2\dot{\theta}^2+2l\dot{x}\dot{\theta}\cos\theta-2l\dot{y}\dot{\theta}\sin\theta\sin\varphi+2l\dot{y}\dot{\varphi}\cos\theta\cos\varphi+l^2\dot{\varphi}^2\cos^2\theta)+\frac{1}{2}M(\dot{x}^2+\dot{y}^2). \quad (2)$$

Then the expression for the potential energy takes the form below:

$$\Pi = mgh = mgl(\cos\theta\cos\varphi+1) \quad (3)$$

As a result of applying the Lagrange procedure, we obtain a system of nonlinear heterogeneous differential equations of the second order unresolved for the highest derivative, describing the motion of a dynamic model in matrix form:

$$M(q(t))\ddot{q}+h(q(t),\dot{q}(t))=Q \quad (4)$$

$$M(q(t))=\begin{bmatrix} M+m & 0 & ml\cos\theta & 0 \\ 0 & M+m & ml\sin\theta\sin\varphi & ml\cos\theta\cos\varphi \\ ml\cos\theta & ml\sin\theta\sin\varphi & ml^2 & 0 \\ 0 & ml\cos\theta\cos\varphi & 0 & ml^2\cos^2\theta \end{bmatrix},\ \ddot{q}=\begin{bmatrix}\ddot{x}\\ \ddot{y}\\ \ddot{\theta}\\ \ddot{\varphi}\end{bmatrix},$$

$$h(q,\dot{q})=\begin{bmatrix}h_1\\ h_2\\ h_3\\ h_4\end{bmatrix}=\begin{bmatrix} -ml\dot{\theta}^2\sin\theta \\ -ml\dot{\theta}^2\cos\theta\sin\varphi-ml\dot{\varphi}^2\cos\theta\sin\varphi-2ml\dot{\theta}\dot{\varphi}\sin\theta\cos\varphi \\ ml^2\dot{\varphi}^2\cos\theta\sin\theta-mgl\sin\theta\cos\varphi \\ -2ml\dot{\theta}\dot{\varphi}\cos\theta\sin\theta-mgl\cos\theta\sin\varphi \end{bmatrix},Q=\begin{bmatrix}Fx\\ Fy\\ 0\\ 0\end{bmatrix}.$$

Where F_x, F_y – are the forces acting on the suspension point of the pendulum along the direction of the axes X, Y, respectively.

2.2. *Linear equations of inverted 2DOF pendulum*

To design control methods, it is necessary to reduce the previously obtained nonlinear equations (4) to linear ones. Let us linearize the equations of our system in upright position, about the point of unstable equilibrium: $\theta=0,\dot{\theta}=0,\varphi=0,\dot{\varphi}=0$. Then we expand our system in a Taylor series in the first approximation. As the initial equations for linearization, a system of nonlinear equations obtained earlier was used. And through linearization we get the following expression:

$$M\ddot{q}+D\dot{q}+Kq=Q \quad (5)$$

We calculate the matrices M, D, K below.

$$M(q(t))=\begin{bmatrix} M+m & 0 & ml & 0 \\ 0 & M+m & 0 & ml \\ ml & 0 & ml^2 & 0 \\ 0 & ml & 0 & ml^2 \end{bmatrix},\ K=\begin{bmatrix} 0 & 0 & 0 & 0 \\ 0 & 0 & 0 & 0 \\ 0 & 0 & -mgl & 0 \\ 0 & 0 & 0 & -mgl \end{bmatrix}.$$

$$D=\begin{bmatrix} 0&0&0&0\\ 0&0&0&0\\ 0&0&0&0\\ 0&0&0&0 \end{bmatrix},\ q=\begin{bmatrix}x\\ y\\ \theta\\ \varphi\end{bmatrix},\ \dot{q}=\begin{bmatrix}\dot{x}\\ \dot{y}\\ \dot{\theta}\\ \dot{\varphi}\end{bmatrix},\ \ddot{q}=\begin{bmatrix}\ddot{x}\\ \ddot{y}\\ \ddot{\theta}\\ \ddot{\varphi}\end{bmatrix}.$$

3. Design of inverted 2DOF pendulum control system

Since the system has split into 2 groups of independent equations along the axes X, Y, respectively, we have the right to synthesize a controller along only one axis. Then, using the equation (6), representation of the system in the state space form was obtained:

$$\begin{bmatrix} 1 & 0 & 0 & 0 \\ 0 & M+m & 0 & ml \\ 0 & 0 & 1 & 0 \\ 0 & ml & 0 & ml^2 \end{bmatrix} \begin{bmatrix} \dot{x}_1 \\ \dot{x}_2 \\ \dot{x}_3 \\ \dot{x}_4 \end{bmatrix} = \begin{bmatrix} 0 & 1 & 0 & 0 \\ 0 & 0 & 0 & 0 \\ 0 & 0 & 0 & 1 \\ 0 & 0 & mgl & 0 \end{bmatrix} \begin{bmatrix} x_1 \\ x_2 \\ x_3 \\ x_4 \end{bmatrix} + \begin{bmatrix} 0 \\ 1 \\ 0 \\ 0 \end{bmatrix} u \quad (6)$$

Now the system of equations in the Cauchy form was represented below to obtain the system in the state space.

$$\begin{bmatrix} \dot{x}_1 \\ \dot{x}_2 \\ \dot{x}_3 \\ \dot{x}_4 \end{bmatrix} = \begin{bmatrix} 0 & 1 & 0 & 0 \\ 0 & 0 & -\frac{mg}{M} & 0 \\ 0 & 0 & 0 & 1 \\ 0 & 0 & \frac{(M+m)g}{Ml} & 0 \end{bmatrix} \begin{bmatrix} x_1 \\ x_2 \\ x_3 \\ x_4 \end{bmatrix} + \begin{bmatrix} 0 \\ \frac{1}{M} \\ 0 \\ -\frac{1}{Ml} \end{bmatrix} u \quad (7)$$

Where $x_1 = x, x_2 = \dot{x}, x_3 = \theta, x_4 = \dot{\theta}, \dot{x}_1 = x_2, \dot{x}_3 = x_4$.

3.1. Modal regulator

Based on the above equations, we obtain the characteristic polynomial $M(s)$ of our system:

$$M(s) = s^4 + (\frac{1}{M}k_2 - \frac{1}{Ml}k_4)s^3 + (\frac{1}{M}k_1 - \frac{(M+m)g}{Ml} - \frac{1}{Ml}k_3)s^2$$

$$+(\frac{mg}{M^2l}k_2 - \frac{(M+m)g}{M^2l}k_2)s + (\frac{mg}{M^2l}k_1 - \frac{(M+m)g}{M^2l}k_1). \quad (8)$$

Then, having specified the necessary arrangement of the system poles from the requirements for the settling time of the transient process and dynamic error, we obtain the desired polynomial and find the value of the matrix of control coefficients. In our case:

$$K = [-361.1167 \quad -180.5584 \quad -487.5520 \quad -92.6490]$$

3.2. Linear-Quadratic regulator

1. To synthesize a linear-quadratic controller, it is necessary to solve the algebraic Riccati equation in the matrix form:

$$P_0 A + A^T P_0 - P_0 B R^{-1} B^T P_0 + Q = 0 \quad (9)$$

The specified transient characteristics are provided by the Q, R matrix values. For our system:

$$K = [-63.2456 \quad -42.7889 \quad -167.6234 \quad -37.2252]$$

4. Simulation of automatic stabilization of inverted 2DOF pendulum

Next, we simulated the stabilization process of inverted 2DOF pendulum by numerical integration of differential equations (4) describing the motion of our system with modal and linear-quadratic regulators. The following initial data were used in the calculation:

$M = 3\ kg,\ m = 0.3\ kg,\ l = 0.3\ m,\ \theta_0 = \varphi_0 = 0.2\ rad,\ x_0 = y_0 = 0$.
Numerical integration of equations (4) was performed in the Matlab environment using the ode15i solver. The simulation results are shown below in Figure 2.

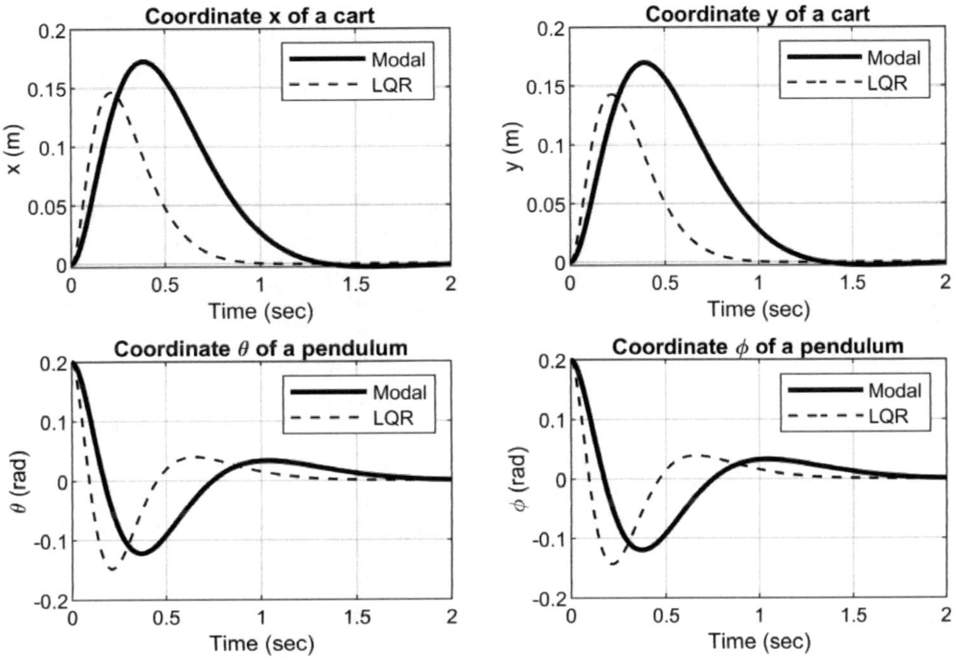

Figure 2. Graphs of the system transient process

From these graphs it can be seen that with a comparable dynamic error value (0.14 rad instead of 0.12 rad), the settling time of the linear-quadratic regulator is much lower (0.5 seconds), i.e. it allows you to achieve a better quality transition process.

5. Conclusion

In this work, a synthesis of controllers was made that allows stabilizing the position of inverted 2DOF pendulum on a cart, since such a control object is an unstable nonlinear system in which some of the degrees of freedom are not directly controlled. Stabilization was carried out using linear controllers with feedback on the state vector: modal and linear-quadratic. To synthesize them, that is, to determine the regulator coefficients, preliminary calculations were carried out to obtain a dynamic model of inverted 2DOF pendulum on a cart. Simulations have shown that the process of stabilization of the model can indeed be carried out using modal and LQ controllers, while the value of the dynamic error in the angle of rotation of the pendulum in the modal controller is rather larger than that of the linear-quadratic controller, and the linear-quadratic controller has significantly better settling time of the transient process, therefore, the LQR regulator copes with the task better.

References

[1] Lim Y Y, Hoo C L and Felicia Wong Y M 2018 Stabilising an Inverted Pendulum with PID Controller *MATEC Web of Conferences* **152** 1–14
[2] Wang J J 2011 Simulation studies of inverted pendulum based on PID controllers *J. Simul. Model. Pract. Theory* **19** 440–49
[3] Patra A K, Mishra A K, Nanda A, Subudhi D K, Agrawal R and Patra A 2020 Stabilizing and trajectory tracking of inverted pendulum based on fractional order PID control *Lecture Notes in Networks and Systems* **109** 338–46

[4] Li W, Ding H and Cheng K 2012 An investigation on the design and performance assessment of double-PID and LQR controllers for the inverted pendulum *Proc. of the 2012 UKACC Int. Conf. on Control, CONTROL* pp 190–6
[5] Prasad L B, Tyagi B and Gupta H O 2012 Modelling & simulation for optimal control of nonlinear inverted pendulum dynamical system using PID controller & LQR *Proc. - 6th Asia Int. Conf. on Mathematical Modelling and Computer Simulation, AMS* pp 138–43
[6] Kumar V and Jerome J 2013 Robust LQR Controller Design for Stabilizing and Trajectory Tracking of Inverted Pendulum *ScienceDirect Int. Conf. On DESIGN AND MANUFACTURING, IConDM Procedia Eng.* 64 pp 169–78
[7] Bräunl T 2003 *Embedded Robotics* (Berlin: Springer)
[8] Mansoor H and Bhutta H A 2016 Genetic algorithm based optimal back stepping controller design for stabilizing inverted pendulum *Proc. Int. Conf. on Computing, Electronic and Electrical Engineering, ICE Cube 2016 – Proc. pp* 1–5
[9] Omatu S and Deris S 1996 Stabilization of inverted pendulum by the genetic algorithm *Proc. of the IEEE Conference on Evolutionary Computation* pp 700–5
[10] Wibowo B C, Subroto I M I and Arifin B 2016 A position controller model on color-based object tracking using fuzzy logic *Int. Conf. on Electrical Engineering, Computer Science and Informatics (EECSI)* 190 pp 1–5
[11] Jain A, Tayal D and Sehgal N 2013 Control of Non-Linear Inverted Pendulum using Fuzzy Logic Controller *Int. J. Comput. Appl.* **69** 7–11
[12] Ray G, Das S K and Tyagi B 2007 Stabilization of inverted pendulum via fuzzy control *J. Inst. Eng. Electr. Eng. Div.* 88 58–62
[13] Tao C W, Taur J S, Wang C M and Chen U S 2008 Fuzzy hierarchical swing-up and sliding position controller for the inverted pendulum-cart system *Fuzzy Sets Syst.* **159** 2763–84
[14] Liu Y, Chen Z, Xue D and Xu X 2009 Real-time controlling of inverted pendulum by fuzzy logic *Proc. of the 2009 IEEE Int. Conf. on Automation and Logistics, ICAL* pp 1180–3
[15] Wu Q H, Hogg B W and Irwin G W 1992 A Neural Network Regulator for Turbogenerators *IEEE Trans. Neural Networks* 3 pp 95–100
[16] Hanwate S D, Budhraja A and Hote Y V. 2016 Improved performance of cart inverted pendulum system using LQR based PID controller and ANN *2015 IEEE UP Section Conf. on Electrical Computer and Electronics, UPCON 2015* pp 1–6
[17] Anderson C W 1989 Learning to Control an Inverted Pendulum Using Neural Networks *IEEE Control Syst. Mag.* 9 pp 31–37
[18] Upadhyay D, Tarun N and Nayak T 2013 ANN based intelligent controller for inverted pendulum system *2013 Students Conf. on Engineering and Systems, SCES 2013* pp 1–6
[19] Artyunin A I, Eliseev S V and Sumenkov O Y 2019 Experimental studies on influence of natural frequencies of oscillations of mechanical system on angular velocity of pendulum on rotating shaft *J. Lecture Notes in Mechanical Engineering* 159–66
[20] Artyunin A I, Barsukov S V and Sumenkov O Y 2020 Peculiarities of Motion of Pendulum on Mechanical System Engine Rotating Shaft *J. Lecture Notes in Mechanical Engineering* 649–57

The design of a support chair for more effective patient rehabilitation

G Mamontov, V Seryakov, Y Khmelevsky and L Tkacheva
Tomsk Polytechnic University, 30, Lenina ave., Tomsk, 634050, Russia

E-mail: hmelevskiy@tpu.ru

Abstract. The authors examined the effect of different color shades on a person and found an effective color and shape solution that can improve the quality of rehabilitation equipment. The paper presents a survey of colors and shapes visual perception by human. As result of theoretical and experimental research, the rehabilitation support chair was designed as a cube-shaped structure with elements painted in shades of dark blue, yellow and white.

1. Introduction

Our contemporary lifestyle is diverse and dynamic. Being outdoors, at work or at home, people consciously or unconsciously contemplate the objects around them. Each of these objects has such parameters as shape and color. What do shape and color of surroundings matter in our lives and how do they influence us? People have become interested in these questions for a long time, and they still remain topical, especially if they concern their professional activities. This paper discusses the choice of a shape and color scheme in the design of a support chair. The support chair is a type of sitting equipment used in rehabilitation of children with cerebral palsy and various congenital disorders. Staying in the support chair is a postural therapy. In order to make this procedure easily tolerable, the equipment is completed with a table which allows a child to draw, write and eat during the rehabilitation.

The main function of the support chair is postural rehabilitation. Its auxiliary functions include assistance in eating and educational activities such as reading, writing, drawing, and various board games. Taking into account these functions, we can argue that the visual appearance of the equipment should create senses of harmony, stability and concentration in order for the rehabilitation to be most effective. This problem can be solved by choosing a shape and color scheme of the product [1].

2. Shape selection

Speaking about the product shape, we should consider static and dynamic properties. Based on the previously defined functions of the support chair, we can say that it should be as static as possible. A square is known to be the most static geometric figure. The volume enhances this property. Therefore, if we take a cube as the basis for designing the product shape, it will be as static as possible from the perspective of visual perception [2]. The concept of statics can also be attributed to one of the basic properties of the subject under study, since its main function is associated with stability, concentration and harmony [3].

To confirm the theory of the greatest degree of static square compared to other figures such as a circle, triangle and rectangle, an experiment was conducted in the form of a survey using Google Forms. In the presented survey, respondents had to distribute points from 1 to 4 and determine how

Content from this work may be used under the terms of the Creative Commons Attribution 3.0 licence. Any further distribution of this work must maintain attribution to the author(s) and the title of the work, journal citation and DOI.
Published under licence by IOP Publishing Ltd

much the selected geometric figures evoke the sensations of harmony, stability, concentration. These sensations are most important for visual perception of the designed product. 84 people participated in the survey. As shown in Table 1, the shape of the square evokes feelings of harmony, stability and concentration of attention more than other figures; the final average value was 10 points.

Table 1. Influence of Shapes on Human Perceptions.

	Square	Circle	Triangle	Rectangular
Harmony	4	4	2	3
Stability	3	1	3	1
Concentration	3	2	1	1
Total	10	7	6	6

Summing up the choice of the shape for the support chair using the analysis of the theoretical data and the experiment, we can argue that a square (or a cube in volume) is the best choice in the design of this equipment that produced senses of stability, harmony and concentration. It should be noted that the circle is ranked the second (7 points on average) and can also be used in the product design. Three-dimensional figures based on a circle, such as a ball or a cylinder, can be used as shapes for minor details of the product design.

2. Choice of color solutions

Now let us consider how different shades of color affect people and how the quality of rehabilitation equipment can be improved using an effective color scheme. To solve these problems, we analyzed some knowledge of colors from the perspective of human visual perception.

Johann Wolfgang Goethe, a German writer, thinker, philosopher and naturalist, was one of the first to become interested in this subject. In his *Theory of Colours*, Goethe argues (cited from [4]): "Since color occupies so important a place in the series of elementary phenomena, filling as it does the limited circle assigned to it with fullest variety... we shall not be surprised to find that it produces an effect on the eye, with which it is immediately associated and by means of which it acts on the mind... producing this impression in their most general elementary character, without relation to the nature or form of the object on whose surface it is apparent". Let us consider how Goethe and other authors describe different colors.

2.1. Yellow

In his *Theory of Colours*, Goethe describes it as follows: "Bright, joyful. Noble and magnificent like gold. Gives an impression of warmth to everything it touches [4]. Max Lusher, a Swiss psychologist, known for inventing a color test [5], describes yellow as the brightest color in the test, perceived as a source of light and cheerfulness. Yellow symbolizes relief from burdens, pressure and problems. In his book *Color Harmony*, Hideaki Chijiwa, a professor of Musashino Art University [6], describes yellow as follows: "Yellow is the radiance of the sun. This lively, cheerful, joyful color is overpowering, and you can get tired of it."

2.2. Orange

Orange or red-yellow. Goethe describes it as follows: "It has the properties of red color, but is not aggressive. It has a very soft and exciting effect, causes a feeling of warmth and joy, promotes a good mood. Orange represents motion, cheerfulness, sociability, achievement, and self-affirmation" [4]. According to Max Lusher, yellow-red is the most stimulating of all colors [5]. If you look at this color for a long time, it will increase your heart rate, blood pressure and breathing. Hideaki Chijiwa "identifies orange as a warm, bright, vibrant and aggressive color, like hot lava erupting from the crater of a volcano" [6]. In his opinion, these hues attract our eyes and evoke emotions to a greater extent than other colors.

2.3. Red

According to Goethe's expression, red is opacity of light [4]. This color can be understood as the most materialized light, reaching its critical point before being lost in darkness. Red is life, energy, passion, will to win, activity, fire, heat, power, impulse, anger, fire, and blood. Lusher writes that the psychological meaning of this color is excitement, activity, and stimulation [5]. A joyful response to red indicates an expression of great strength, stimulation, and self-confidence. Red represents the "appetite", physical activities like sports and sex, emotional excitement, passion, and ardor. Feeling antipathy towards red color, a person perceives it as a source of threat, anguish, disgust, and anger. In his book *Color Harmony*, Hideaki Chijiwa notes that red is full of passion, that it is the color of heart and fire [6].

2.4. Violet

Johann Wolfgang Goethe calls this color "blue-red" and mentions the Catholic higher dignitaries who have appropriated this unquiet color to themselves; anxiety is growing and striving for cardinal purple [4]. Violet is full of dignity; it is calm, and restrained. People see something serious, mystical, mysterious, and heavy in this color. It can even lead to depression. There is an opinion that the name of the color comes from the Spanish word "violar: which means "to dirty", "to stain", "to spoil". It can be assumed that the viola, a low-register string instrument, producing a solemn, mystical sound, was named so because it was associated with the darkness that would come beyond the violet. Max Lusher called violet the color of transcendence, the transition to a world dominated by emotional values, as well as a representation of impressionability, enthusiasm, curiosity, and identification [5]. Hideaki Chijiwa associates this color with a regal appearance and calls it somewhat "artificial" since it is quite rare in nature [6].

2.5. Light blue

Goethe says: "As yellow is always accompanied with light, so it may be said that blue still brings a principle of darkness with it. "As a hue it is powerful, but it is on the negative side, and in its highest purity is, as it were, a stimulating negation. Its appearance, then, is a kind of contradiction between excitement and repose" [4]. Light blue represents the sky, ice, cold, isolation, tact, and sincerity. This color relieves stress, calms, promotes the development of imagination, liberates, and leads away from reality. Hideaki Chijiwa describes light blue as youthful and athletic [6].

2.6. Dark blue

Johann Wolfgang Goethe says that "blue gives us an impression of cold, and thus, again, reminds us of shade" [4]. Blue represents wisdom, depth, calmness, vision, silence, idealism, constancy, devotion, seriousness, dedication, and severity. One of the most famous poems by I. V. Goethe fully captures the essence of dark blue:

"Over the tops of the mountains in peace;
In the trees scarce a breath stirs their crest;
And the birds in the wood singing cease;
Only wait – soon though too shalt have rest."

It attracts and fascinates, invites us to meditation, to searches for life's purpose and truth, but it can lead to melancholy and weakness of spirit. Max Lusher writes that blue evokes a sense of the deepest peace, relaxation, satisfaction, harmony, orderliness, unity, and security. It is no coincidence that the veil of the Virgin Mary is traditionally painted in blue. Hideaki Chijiwa described blue as a color creating an atmosphere of well-being [6].

2.7. Green

Goethe calls green the color of satisfaction, tranquility and balance [4]. Green is nature and harmony. It soothes and pacifies, promotes concentration, calmness and serenity, helps to concentrate and make a decision. Moreover, green contains potential energy, contributes to prosperity and stability. The Russian word "green" (*zeleny*) comes from the Old Russian word "*zelo*" which means "highly", "greatly", "very", "strongly". Lusher characterizes green as creating stability and constancy, emphasizing a sense of self-worth, expressing as respect, dignity, authority, and competence [5]. Hideaki Chijiwa says that green reminds us of spring and fertility, but it is also the color of mold, poison and jealousy [6].

2.8. Brown
This color is formed by mixing red with black, which makes it more "earthy", expressing constancy and reality. Calm, strong earthy hues of brown give an impression of something massive, confident and compact [4]. Lusher writes that brownish red has a connotation of peace [5]. Hideaki Chijiwa says that brown is rich, "soil-like", but also sad like autumn leaves [6].

2.9. White
Being a substitute for light, white makes our eye work [4]. Lusher writes that white is the lightest of all colors [5]. It can be perceived as dazzling. White is the color of purity and innocence. In Christian traditions, it is the color of holiness. Hideaki Chijiwa describes white as purity, virginity and peacefulness. On the other hand, it represents winter and hospital sterility [6]. Attributes of Colors are presented in table 2.

Summing up the information on human visual perception of color, we can argue that such colors as dark blue, light blue, green and white contribute to concentration and produce a sense of stability and harmony [7]. Yellow, orange and red evoke such emotions as cheerfulness, joy, and vigor, help to improve mood, and increase the motivation for rehabilitation sessions [8]. This condition is necessary for effective rehabilitation, since people with serious illnesses are often depressed and lose the desire to systematically conduct exercise sessions using rehabilitation equipment.

The authors carried out a survey on human perception of color shades, which can evoke a feeling of harmony, stability and concentration. In the survey, respondents had to distribute points from 1 to 9 and determine how different shades evoke the sensations mentioned above. Based on the main function of the support chair - fixing the patient in the sitting position, a color solution creating a feeling of concentration, harmony and stability was determined. The survey involved 82 people of different sex and age. As shown in Table 3, dark blue, with an average score of 24 points, produces senses of harmony, stability and concentration to the greatest extent.

Table 2. Attributes of Colors.

	Goethe	Max Lusher	Hideaki Chijiwa
Yellow	Clear, joyful, noble	Bright, cheerful, liberating	Sunny, lively, cheerful, joyful
Orange	Joyful, warm, conducive to good mood	Energizing	Warm, bright, colorful, aggressive
Red	Serious, full of dignity, charming, passionate, pretty	Exciting, energizing, stimulating, strengthening, empowering, appetizing, sexy, passionate, angry, tormenting	Passionate, hearty, fiery
Violet	Disturbing, insufferable in its dark shades	Impressive, enthusiastic, curious	Regal, unrealistic

Light blue	Combining excitement and peace, running away from a perceiver, cold, shady	-	Youthful, athletic
Dark blue	Carrying you to a place where you can rest. Lively but joyless	Calm, relaxing, satisfying, harmonious, protective	Creating an atmosphere of well-being
Green	Satisfying, soothing, balancing	Stable, consistent, meaningful, authoritative, competent	Spring, fertile, poisonous, jealous
Brown	Unpleasant, shameful, disgusting.	Pacifying	Rich, earthy, sad
White	-	The lightest, dazzling, pure, innocent, holy	Pure, virgin, peaceful, wintry, sterile

Table 3. Influence of Colors on Human Perceptions.

Color	Concentration	Harmony	Stability	Total
Yellow	3	4	4	11
Orange	2	2	3	7
Red	1	1	2	4
Violet	4	3	1	8
Light blue	5	6	5	16
Dark blue	8	7	9	24
Green	6	7	8	21
Brown	8	4	6	18
White	7	8	7	22

3. Conclusion

Autodesk Fusion 360 was used to develop a 3D model of a support chair based on theoretical and experimental studies. The general shape of the support chair is presented in the form of a cube, and the functional elements were made in the form of other forms using basic colors: blue for stability, yellow for cheer, and white for purity. Based on the 3D model of the chair-support, it is planned to register an industrial model and get a patent for the invention.

The image of the designed equipment is shown in Figure 1.

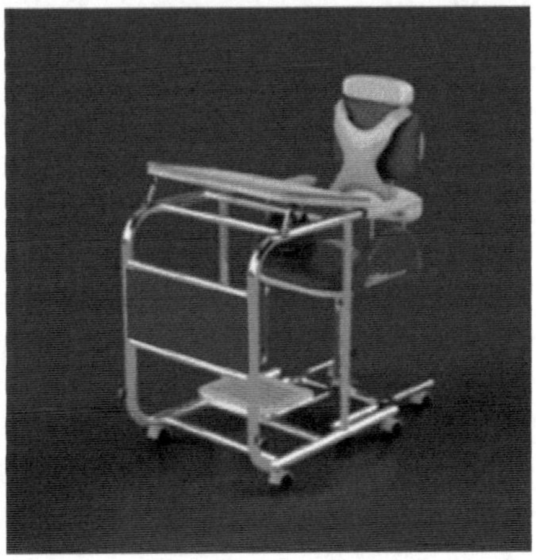

Figure 1. Image of the rehabilitation support chair.

References

[1] Kochegarov B E 2006 *Industrial Design* (Vladivostok: Far Eastern State Technical University)
[2] Mikheeva M M 2008 *Color in Design* (Moscow: Bauman Moscow State Technical University),
[3] Lusher M 1997 *Color of Your Character* (Moscow: Veche)
[4] Chijiwa H 2003 *Color Harmony: A Guide to Creative Color Combinations* (Moscow: Astrel Publishing House LLC)
[5] Seryakov V A, Khmelevsky Y P and Mamontov G Y 2015 The Design Features of the Body of the Portable Electrocardiograph "ECG-EXPRESS" *IOP Conf. Series: Materials Science and Engineering* **93** 012076
[6] Medvedev V Y 2005 *Color Science* (St. Petersburg: Publishing and Printing Center of St. Petersburg University of Technology and Design)
[7] Khmelevsky Y P 2017 A Study of Shaping for the Body of a Rehabilitation Exercise Machine *20th National Scientific and Practical Conf. Technology of Artistic Material Processing* (Rostov-on-Don: Don State Technical University)

Quality Ratios for Public Transport Services

A N Novikov[1], S V Eremin[2], A G Shevtsova[3], D O Lomakin[1] and A V Kulev[1]
[1]Orel State University, 95, Komsomolskaya st., Orel, 302026 Russian Federation
[2]Krasnoyarsk City Administration, 93, Karl Marks str., Krasnoyarsk, 660049, Russian Federation
[3]Belgorod State Technological University named after V.G. Shukhov, Kostyukova str., Belgorod, 308012 Russian Federation

Email: shevcova-anastasiya@mail.ru

Abstract. The system of quality indicators used for public transportation is proposed by the authors. Additionally, the paper considers the necessity to develop an assessment ratio, which defines a competitiveness level among enterprises. To assess the activities of a transport operator, the authors proposed to apply a four-level system that will assess the quality of transportation: the level of service may be "excellent", "good", "satisfactory" and "unacceptable".

1. Introduction
When transporting cargo, the main criteria for quality are delivery speed and cargo safety. If these two things are fulfilled in accordance with customers' requirements, then the level of quality can be rated as high. At the same time, if some discrepancies between the delivery speed and cargo safety indicators are observed as they have been originally declared in the contract, the transport service provider (operator) reimburses a customer for the costs paid for these services. So, for example, in case of damage or shortage of cargo during transportation, an operator pays the cost of the missing or damaged cargo. And if the delivery time is exceeded by the fault of an operator, he/she must pay the fines in the amount that has been previously agreed in the contract between the parties.
When carrying out passengers' transportation, the end consumer of this service is a passenger. When traveling by public transport, the main quality criteria will be: the time spent on moving and the level of comfort provided in this trip. If we make an analogy with cargo transportation, then these two quality indicators are quite close by their nature: in both cases, the end consumer is concerned, by and large, only two parameters – delivery speed and safety [1]. When transporting cargo, a consumer is concerned about the cargo safety, and when taking the public passenger transport he/she expects health safety.

2. Ratio system used to assess quality of public transport services
To express the intangible indicators of transportation quality through numerical equivalents, the ratio system is used aimed at assessing the quality of public transport services [2]. The ratio (K_f) shows how full a passenger's cabin of a road vehicle is. The ratio of relative filling is used when assessing the quality of passengers' services, and defined as the standardized ration (Y_s) to the actual one (Y_a).

$$K_f = \frac{Y_S}{Ya} \quad (1)$$

Content from this work may be used under the terms of the Creative Commons Attribution 3.0 licence. Any further distribution of this work must maintain attribution to the author(s) and the title of the work, journal citation and DOI.
Published under licence by IOP Publishing Ltd

In calculations, the ratio value can be greater than unity, because the actual filling of the passenger's cabin in "pre-rush", "inter-rush" and "post-rush" time is less than the standard. In such cases, the filling factor is taken equal to unity (1). The ratio (K_r) shows the regularity of vehicles movement on the route.

$$K_r = \frac{n_i}{N_i} \tag{2}$$

For intra-city transportation, when calculating the ratio K_r, the values of the numerator and denominator correspond to:

n_i - the number of intervals in traffic that meet the standard indicators, within the permissible deviations;

N_i – the total number of traffic intervals.

For suburban transportation, when calculating ratio K_r, the values of a numerator and denominator correspond to:

n_i - the actual number of arrivals and departures of road vehicles to the final and intermediate stop-points according to the established schedule within the permissible deviations;

N_i - the total number of arrivals and departures of road vehicles.

For suburban transportation, when calculating the coefficient K_r the values of the numerator and denominator indicators correspond to: n_i - the actual number of arrivals and departures of road vehicles to the final and intermediate stopping points according to the established schedule, within the limits of permissible deviations;

N_i - the total number of arrivals and departures of vehicles.

Ratio K_r must be different for urban and suburban road vehicles, because not for every urban route there is a schedule. It is convenient to determine the traffic regularity for suburban road vehicles exactly according to the traffic schedule, because the intervals on these routes are usually long. The value of the permissible deviation in this ratio is taken equal to ± 2 minutes.

The ratio of relative time spent on a trip (K_t) is defined as the ratio of the time spent on the trip in the established standardized comfortable conditions (T_s) to the real time spent on the trip (T_r).

$$K_t = \frac{T_s}{T_r} \tag{3}$$

According to the developed standard offered by the State Unitary Enterprise "Mosgortrans", the standardized time for a passenger on the way is 27 minutes. Service organization ratio (K_o) - which shows the actual level of organizational service is defined for a specific route by the number of proposals and complaints for a certain time.

$$K_o = 1 - \frac{n}{N} \tag{4}$$

where n - is the number of complaints and proposals for a certain time; N - the total number of passengers transported during this time period.

Complaints and proposals from passengers can be received both through telephone or a messenger, the numbers of which are in the passenger cabin of a road vehicle, and by interviewing passengers through a comprehensive survey. Only complaints from passengers are taken into account, which cannot be taken into account in the existing ratios. It means that the complaints regarding regularity violations towards the vehicles' moving, the excess filling of a road vehicle cabin, failure to provide the transport services, etc. are not taken into account, since these violations are already reflected in other ratios [3].

In a number of studies on this topic, a different approach was used to define the value of this ratio [4] The Passenger Motor Transport Company carried out an examination and identified some violations that occurred on the route. The indicator n was the number of movements performed with a violation and indicator N was the total number of movements done on this route. The inconsistency of the approach consisted in the fact that a motor transport company was the executing and controlling body. Under the new interpretation of the formula, the "controlling authority" is passengers.

Traffic safety ratio (K_{ts}) – the level of transportation quality is characterized to a large extent by traffic safety.

$$K_{ts} = \frac{1}{1 + A_{tl} B_a} \tag{5}$$

where A_{tl} - time losses ratio associated with road accidents; B_a – the number of accidents caused by a driver's fault per 1 million km, where B_a is calculated with formula:

$$B_a = \frac{n}{L} \tag{6}$$

where n – is the number of accidents caused by a driver of a bus (trolleybus, tram); L – the total mileage of vehicle stock in the current year, million km;

However, when the system for measuring the degree of a driver's fault in penalty points has been cancelled, such a formula became irrelevant. Therefore, it is proposed to define the dynamic indicator showing the level of road accidents at an enterprise based on the statistics of the number of violations for a certain period of time, regardless of their severity.

The ratio of non-provision of transport services (K_n) is defined as:

$$K_n = \frac{Q_n}{Q_n + Q_{np}} \tag{7}$$

where Q_n is the number of passengers who use the transport service;

Q_{np} - the number of passengers who did not use the transport service for various reasons (inability to board the vehicle, use of an alternative travel option (taxi, route taxi, walking)).

The analysis among research and practical works within this topic has showed that the ratio for assessing the quality level regarding passengers' motor transport enterprises has not been used before.

3. Integral quality assessment for passengers transport services

An integral assessment of quality of transport services for passengers has already been presented in different ways in a number of works addressing this topic. However, as a rule, the applied methodology did not take into account the fact that transport operators have to work in the competitive environment. For the correct assessment applied to evaluate the quality level, it is necessary to introduce another ratio that would demonstrate the level of competitiveness for a passenger-motor-transport enterprise. This ratio is the fact of transport service non-provision (K_n).

Table 1. Ratio standards for transport services quality

Service level	Standards for quality ratios						
	K_f	K_t	K_r	K_o	K_{ts}	K_n	K_{sq}
Unacceptable	Below 0.77	Below 0.81	Below 0.95	Below 0.91	Below 0.81	Below 0.84	Below 0.5
Satisfactory	0.86	0.86	0.95	0.92	0.84	0.91	0.5
Good	0.93	0.93	0.96	0.96	0.93	0.94	0.7
Excellent	1	1	0.99	0.99	0.99	1	0.95

The integral assessment of quality of transport services for passengers is calculated by the total impact on the service quality ratio (K_{sq}) relating to the vehicle filling (K_f), the regularity of vehicles traffic on the route (K_p), the passenger's time spent on the trip (K_t), the general organization of service (K_o), and traffic safety when transporting passengers (K_{ts}) and total demand response to passenger transport services (K_n):

IOP Conf. Series: Materials Science and Engineering 1118 (2021) 012019 doi:10.1088/1757-899X/1118/1/012019

$$K_{sq} = K_f \cdot K_r \cdot K_t \cdot K_o \cdot K_{ts} \cdot K_n \quad (8)$$

To assess the activities of a transport operator, it is proposed to apply a four-level system that will assess the quality of transportation: the level of service may be "excellent", "good", "satisfactory" and "unacceptable".

According to the results of the survey, the activities of the transport operator belong to 1 out of 4 quality levels, and certain actions are undertaken within the contract concluded between the customer of transportation and the route operator:

- unacceptable level of quality – the contract for transportation services is broken or not extended with the operator; if the operator undertakes to correct the deficiencies within the prescribed time, then after an expert meeting, only a large fine may be imposed on the operator.
- satisfactory level of quality – the operator is issued a warning of the recommendation nature to improve the quality of transportation; the contract is concluded (extended) with the term that the operator will reach a higher quality level by a certain time; also the operator may be fined.
- good quality level – the contract for carrying out transportation activities on the route is extended.
- excellent level of quality – upon reaching this level, the operator is paid a bonus in the amount specified in the terms of the concluded contract.

4. Conclusions

The study resulted in establishing the main values of indicators used to assess the quality of public transport; the indicators' system has been proposed in which the ratio enabling to identify the competitiveness level of a passenger-motor-transport enterprise has been defined.

References

[1] Evtiukov S, Karelina M and Terentyev A 2018 A method for multi-criteria evaluation of complex safety characteristic of road vehicle. *Transportation Research Procedia* **36** 149–156

[2] Novikov A, Novikov I and Shevtsova A 2019. Modeling of traffic-light signalization depending on the quality of traffic flow in the city. *Journal of Applied Engineering Science* **17(2)** 175–181

[3] Marusin A and Danilov A 2018 Method for assessing influence of automated traffic enforcement system parameters on traffic safety. *Transportation Research Procedia* **36** 500–506.

[4] Gorev A, Solodkiy A and Enokaev V 2018 Improving efficiency of traffic management and safety based on integration of local ATMS. *Transportation Research Procedia* **36** 207–212.

[5] Novikov A, Novikov I, Katunin A and Shevtsova A 2017 Research of influence of dynamic characteristics for options controlled intersection. *Procedia Engineering* **187** 664 – 671

[6] Glagolev S, Shevtsova A & Shekhovtsova S 2018. Basis for application of new-generation anti-icing materials as an efficient way to reduce the accident rate on roads in winter. *Transportation Research Procedia* **36** 193–198

[7] Kapitanov V 2018 Methods for traffic management efficiency improvement in cities. *Transportation Research Procedia* **36** 252–259

Path-building and localization system for mobile robotic platforms based on Raspberry Pi

E A Ivanov, A S Belyaev and O A Brylev

Tomsk Polytechnic University, 30, Lenin Avenue, Tomsk 634050, Russia

E-mail: eai13@tpu.ru

Abstract. The paper demonstrates a system for detection of location of robotic platform FESTO Robotino and optimal route building. It processes data from the camera and transmits control signals to the control system of the robot. The whole system is based on Raspberry Pi. It detects robot's current coordinates, current angular rotation, angular difference (difference between current and previous angular rotation) and displacement of the robot in its own coordinate system. It uses an ArUco marker, placed on the top of the mobile robot for that. System also builds an optimal path, when moving from one point of the surface to another, according to the permeability of the surface. The authors set the permeability of testing surfaces. Using that, a weighted graph is built through the centers of particular surfaces, which are detected via an algorithm on Raspberry Pi. The optimal path is constructed through the edges of the graph via modified Dijkstra algorithm.

1. Introduction

With technological and industrial development, modern mobile autonomous robotic systems gradually move from laboratories to the harsh environmental conditions. Predicting the environmental impact on the system, in this case, is usually impossible. It makes control systems for mobile robots become quite complicated. Among the basic mobile robotics' problems, the navigation task is still actual. It includes detecting current position of the robot via getting the data from the on-board sensor system, and path-planning, according to the permeability of the surface. The latter also consists of the optimal route building, obstacle-avoiding algorithm and others [1-5]. For this purpose, a system was constructed. It includes a testing surface [6-7], imitating the heterogeneity of the ground, a camera, placed right above that surface, and a single-board computer Raspberry Pi. The computer processes the image from the camera and the data from the robotic platform, and pushes signals to the mobile robot, moving on the stand. The structural scheme of the system and the image of the testing ground are presented in Figure 1.

The authors of [6-7] implemented a system for robot's position and angular rotation detection. Proposed system is also used for surface recognition, according to the data from the on-board sensor system. Specially created computer vision system, mounted above the the testing surface, is used for training and for evaluation of the results. It detects the location of the robot via the LEDs, set on it. However, such system is not sustainable, when there is no constant luminance. So, one of the objectives of current work is improvement of the location and angular rotation detection. The detection will be based on Raspberry Pi. It is enough for processing the image and all the algorithms. In addition, it is small enough for comfortable integration. The algorithm, detecting current location

Content from this work may be used under the terms of the Creative Commons Attribution 3.0 licence. Any further distribution of this work must maintain attribution to the author(s) and the title of the work, journal citation and DOI.
Published under licence by IOP Publishing Ltd

and angular rotation, includes linear correction of brightness in the picture, detection of the ArUco marker and conversion of coordinates of the ArUco marker into the robot's local coordinate system.

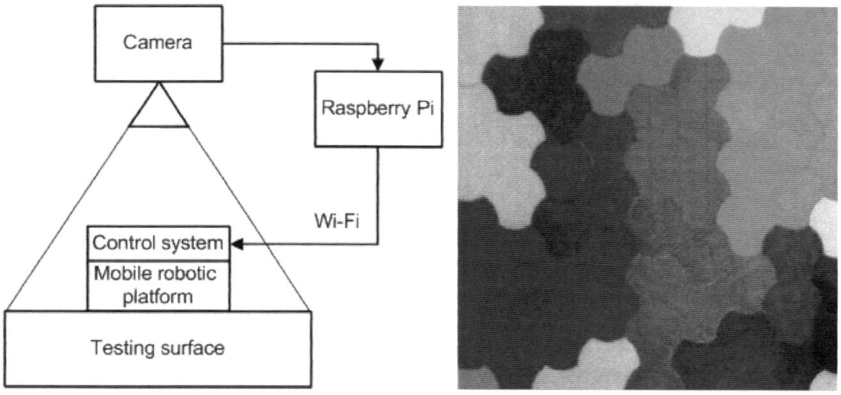

Figure 1. Structural scheme and image of testing surface.

2. Coordinates and angular rotation detection

In order to detect the location of the mobile robotic platform, a special ArUco marker was mounted on its upper side. A typical ArUco marker is demonstrated in Figure 2.

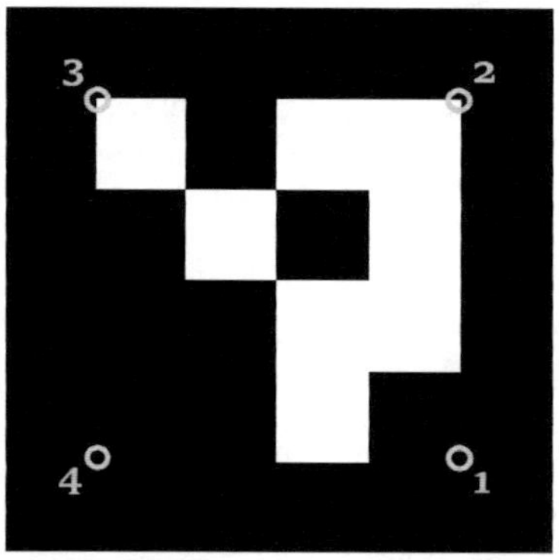

Figure 2. ArUco marker with detected corners highlighted with circles.

For ArUco marker detection a linear correction of the image was made. After that, a built-in function of OpenCV was executed for detecting four corners of ArUco marker. For proper robot controlling a set of data must be received: current location relative to the camera (in pixels), angular rotation relative to the camera, angular difference (difference between current and previous angle rotation), displacement relative to robot's local coordinate system (in millimeters). Having the coordinates of the ArUco corners, and using formula 1, we obtain the location of the center of the robot.

$$x = \frac{x_1 + x_3}{2}, \ y = \frac{y_1 + y_3}{2}, \tag{1}$$

where x_1, x_3, y_1, y_3 – coordinates of two diagonally oriented ArUco corners, x, y – ArUco marker center coordinates.

According to the orientation of the robot's local coordinate system, demonstrated in Figure 3, which was set as zero angular deviation, its evaluation is processed with formula 2.

Figure 3. Coordinate system of FESTO Robotino.

$$\alpha = \begin{cases} \arccos \dfrac{x_2 - x_3}{\sqrt{(x_2 - x_3)^2 + (y_2 - y_3)^2}}, \alpha \in [0; \pi), \\ \arccos \dfrac{x_2 - x_3}{\sqrt{(x_2 - x_3)^2 + (y_2 - y_3)^2}} + \pi, \alpha \in [\pi; 2\pi), \end{cases} \tag{2}$$

where x_2, x_3, y_2, y_3 – coordinates of two ArUco corners, forming the line segment, crossing the positive half-axis of OX.

Angular difference is obtained using formula 3.

$$\Delta\alpha = \alpha(t) - \alpha(t - \Delta t), \tag{3}$$

where $\alpha(t)$ – current angular rotation, Δt – sampling time.

Local displacement is obtained using formula 4.

$$\begin{pmatrix} \Delta x_{local} \\ \Delta y_{local} \end{pmatrix} = K \cdot \begin{pmatrix} -\sin\alpha & -\cos\alpha \\ -\cos\alpha & \sin\alpha \end{pmatrix} \cdot \begin{pmatrix} \Delta x \\ \Delta y \end{pmatrix}, \tag{4}$$

where $\Delta x, \Delta y$ – displacements of the robot relative to camera, $\Delta x_{local}, \Delta y_{local}$ – displacements of the robot in its local coordinate system, K – coefficient of conversion from pixels to millimeters.

In order to convert local displacement value from pixels to millimeters, the coefficient K is defined. It is calculated as the length of the side of ArUco marker in real, divided by its length in the image.

The above algorithm computes the location of the robot in global coordinate system, its angular rotation and displacement in local coordinate system.

3. Path-building algorithm implementation

Authors of [5-6] analyzed robot's movement across various types of surfaces. Based on that, the permeability of testing surface is evaluated. Each surface has its own amount of labor (in points of labor per pixel): Light-grey – 1, Pink and Light-green – 2, Green – 3, Beige, Blue and Dark-green – 4, Grey and Brown – 5, Red – 10.

Segmentation algorithm was implemented for detecting the centers of the surfaces. Firstly, the image is converted from RGB model to HSV model. Then, for each type of surface, threshold binarization is used. As a result, each surface is highlighted. A set of bit images for testing surface is represented in Figure 4.

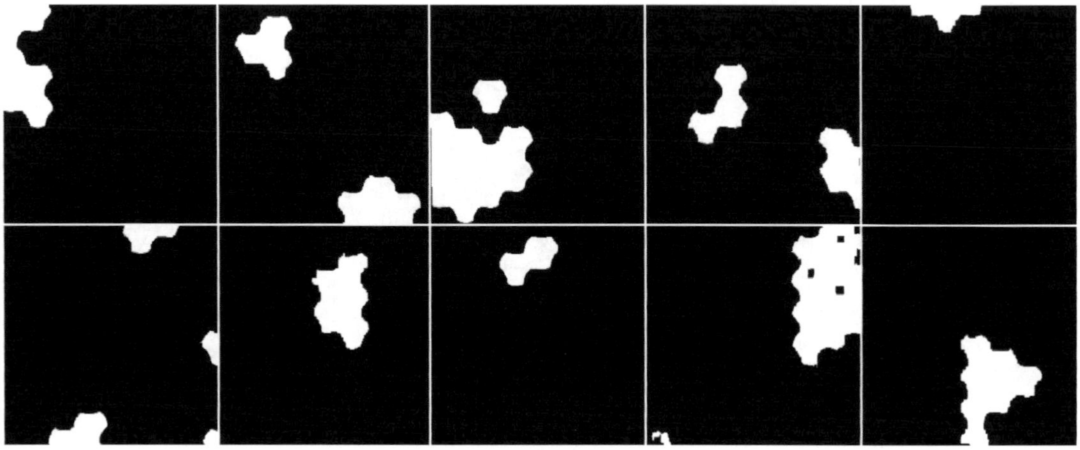

Figure 4. Set of bit images with highlighted surfaces.

Such algorithm is not among the most reliable, but, since the research is held in laboratory conditions, this method satisfies all the needs. However, it will be refined further. Next step is detection of the centers of separated surfaces. It is achieved via determination of mass center for each indispensable surface. Centers of surfaces are obtained using formula 5.

$$x_{c_i} = \frac{\sum_{j=1}^{n_i} x_{ij}}{n_i}, y_{c_i} = \frac{\sum_{j=1}^{n_i} y_{ij}}{n_i}, \quad (5)$$

where x_{c_i}, y_{c_i} – coordinates of the center of the *i*-th surface, n_i – amount of pixels of the *i*-th surface, x_{ij}, y_{ij} – coordinates of pixels of the *i*-th surface.

Then a weighted graph is constructed through the centers of each surface. In order to exclude the mutual influence of different types of surfaces, the limitations are introduced. All of the edges of the graph, except those, which cross only two surfaces, are removed. The final graph is demonstrated in Figure 5. This graph is weighed, according to the amount of points of labor of each surface. That is completed via counting the sum of point of labor for each pixel of particular edge of the graph.

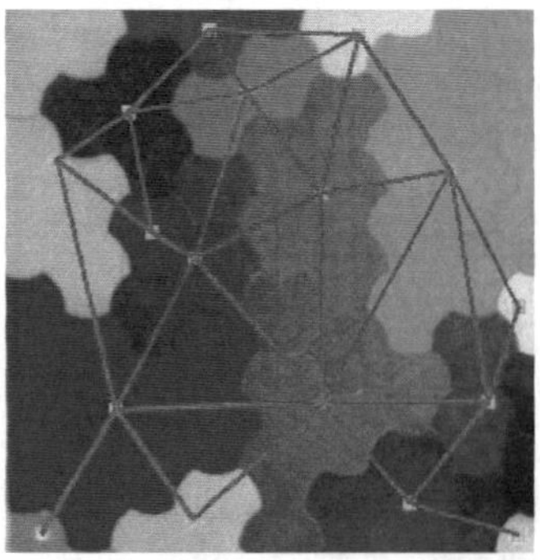

Figure 5. The final graph with limitations introduced.

Now, as the graph becomes weighted, the optimal path building algorithm is implemented. Dijkstra algorithm is used for that. It looks for the least cost route, when moving from one particular surface center to another. But this method defines only the amount of points of labor for the optimal path, but does not image the whole route. So, the algorithm was improved, so that it writes the sequence of the edges of the graph in the route. The whole Dijkstra algorithm and the supplement for it were implemented in C++. Its block diagram is presented in Figure 6.

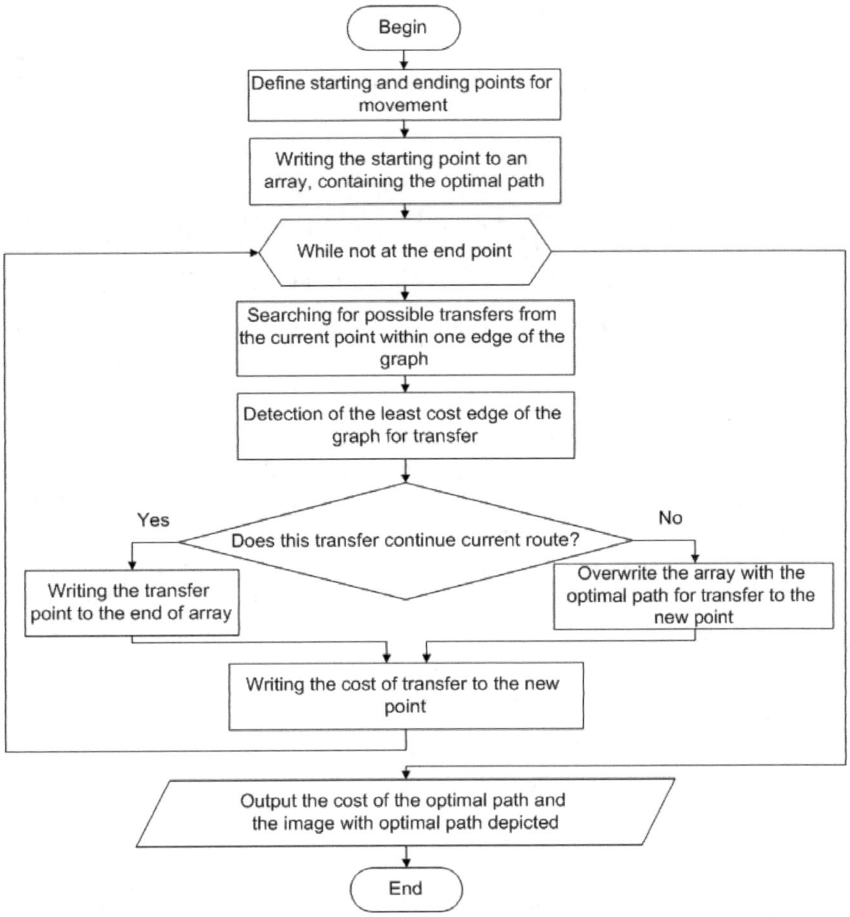

Figure 6. Block-diagram of optimal path building algorithm.

As a result, the algorithm builds the optimal route. Examples of optimal routes are presented in Figure 7. Red routes are optimal, blue ones are the shortest (include the least possible pixels amount).

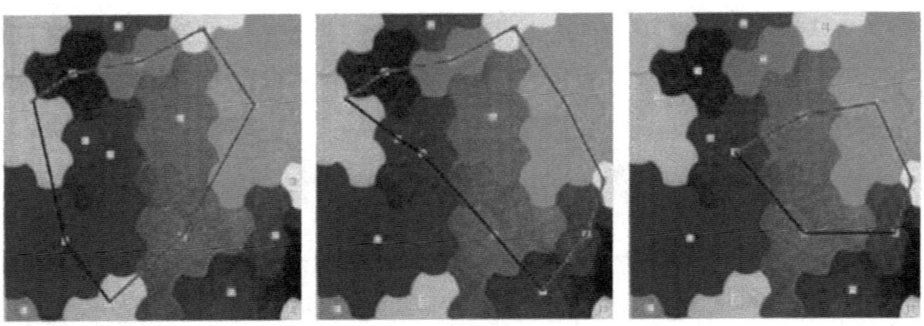

a) b) c)

Figure 7. A set of examples of optimal (red) and shortest (blue) routes.

Figure 7a shows the optimal route with 1360 labor points and 617 pixels of length, while the shortest (blue) path takes 1634 labor points and 259 pixels of length.

In figure 7b, the optimal path takes 1308 points of labor and 578 pixels of length, and the shortest route is 1387 points of labor and 325 pixels of length.

Figure 7c represents the optimal route with 843 points of labor, 346 pixels of length, and the shortest route with 1015 points of labor, 236 pixels of length.

As it appears, the shortest path may not be the optimal one. That is an important aspect to be analyzed, as it influences the speed of robot's movement from one point to another.

4. Conclusion

As a result, this work provides an algorithm for coordinates and angular orientation detection, based on a computer vision system, based on Raspberry Pi. Optimal path building algorithm, which generates the least-labor route between two points, was also implemented. This whole system will further be used solving various navigation tasks and carrying out further investigations, related to mobile robotic platforms, moving across harsh surfaces.

References

[1] Howard A, Seraji H and Werger B 2002 Fuzzy terrain-based path planning for planetary rovers *IEEE Int. Conf. on Fuzzy Systems* pp 316-20

[2] Zhou L, Yang L and Tang H 2017 Research on path planning algorithm and its application based on terrain slope for slipping prediction in complex terrain environment *Int. Conf. on Security, Pattern Analysis, and Cybernetics, SPAC 2017* pp 224-7

[3] Iwasa A, Toda Y and Kubota N 2018 The return way path planning of an autonomous mobile robot considering traveling risk of the road *Proc. – Joint 10th Int. Conf. on Soft Computing and Intelligent Systems and 19th Int. Symp. on Anvanced Intelligent Systems, SCIS-ISIS 2018* pp 1406-9

[4] Guo Y, Song A, Bao J and Zhang H 2011 Optimal path planning in field based on traversability prediction for mobile robot *Int. Conf. on Electric Information and Control Engineering, ICEICE 2011 – Proc.*

[5] Tyryshkin A and Andrakhanov A 2009 Application of GMDH algorithms in the obstacle recognition problem for autonomous mobile robots *Pattern Recognit. Image Anal. 19 2009* pp 197-203

[6] Andrakhnov A and Belyaev A 2017 Navigation learning system for mobile robot in heterogeneous environment: Inductive modeling approach *Proc. of the 12th Int. Scientific and Technical Conf. on Computer Sciences and Information Technologies, CSIT 2017* pp 543-8

[7] Andrakhnov A and Belyaev A 2018 GMDH Based Learning System for Mobile Robot Navigation in Heterogeneous Environment *Advances in Intelligent Systems and Computing* vol 689 ed N Shakhovska, V Stepashko (Cham, Switzerland: Springer) pp 1-20

Evaluation of the possibility of using multilayer nanostructured coatings to improve the output characteristics of roller screw mechanisms

A V Zhdanov, E A Novikova, N S Dovbysh and A N Mitrofanov

Vladimir State University, Mechanical Engineering Technology Department, 87 Gorky Street, Russia

E-mail: zhdanov@vlsu.ru

Abstract. This article examines the effect of different coatings made of different materials on the efficiency of roller gears. Graphs of the dependences of the parameters of each coating are plotted. Tables of changes in the efficiency parameters and friction coefficients for the studied materials are given.

1. Introduction

Reducing the efficiency of gears due to friction is one of the most important problems in mechanical engineering. Various methods of solving this disadvantage are widely used. In particular, a change in the design of the actuator. In this paper, the analysis of changes in the value of the efficiency of RVM is carried out by applying nanostructured coatings to the working surfaces.

2. Experimental study of the properties of nanostructured coatings of roller screw mechanisms rollers

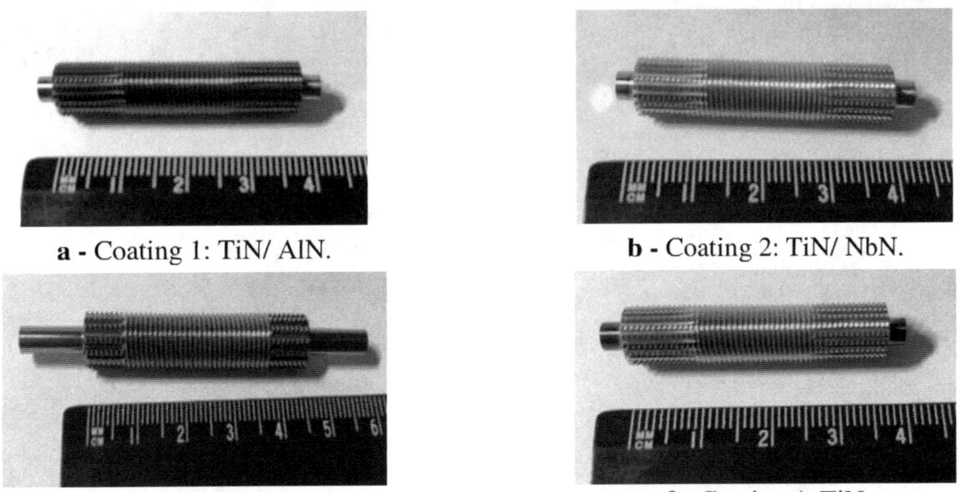

a - Coating 1: TiN/ AlN. **b** - Coating 2: TiN/ NbN.

c - Coating 3: CrN/ CNx. **d** - Coating 4: TiN.

Figure 1. Photos of roller samples with different types of coatings.

Content from this work may be used under the terms of the Creative Commons Attribution 3.0 licence. Any further distribution of this work must maintain attribution to the author(s) and the title of the work, journal citation and DOI.
Published under licence by IOP Publishing Ltd

IOP Conf. Series: Materials Science and Engineering 1118 (2021) 012021 doi:10.1088/1757-899X/1118/1/012021

During the project, four types of nanostructured coatings for rollers were obtained (figure1, a-d):
- Coating 1: TiN/ AlN;
- Coating 2: TiN/ NbN;
- Coating 3: CrN/ CNx;
- Coating 4: TiN.

Tribometrical parameters were taken in the following modes of operation on the CSM tribometer: radius-6.84 mm; ball radius – 5.5 mm; linear speed – 20 cm/s; normal load – 5 N, path – 500 m; temperature – 25°C, humidity 60%, numerical results of scratch testing - in table 1[1].For all six types of coatings, physical and mechanical properties were studied using a microcombitester and a CSM tribometer, the research methods of which were developed at the 1-st stage of the project. The results of measurements of the physical and mechanical properties of coatings are shown in table 1.

Table 1. Results of measurements of physical, mechanical and tribological properties of nanostructured coatings.

	H (GPa)	*HV*	*E (GPa)*	H^3/E^2	μ	*H (μm)*
Coating 3 TiN/AlN	57	1239	319	1.84	0.5	2.0
Coating 4 TiN/NbN	54	2554	345	1.32	0.69	3.05
Coating 5 CrN/CNx	43	2624	229	1.53	0.15	2.5

Figure 2 presents one of many coating photographs after scratch test, tribometrical and thickness measurements [2].

a – Thickness h = 3.053 μm.

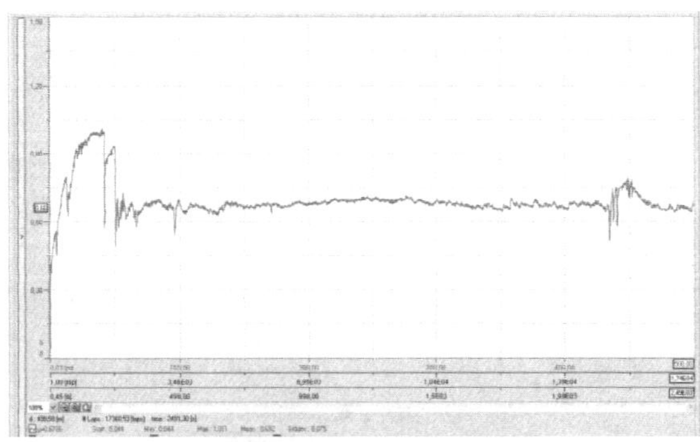

b – Coefficient of friction 0.69.

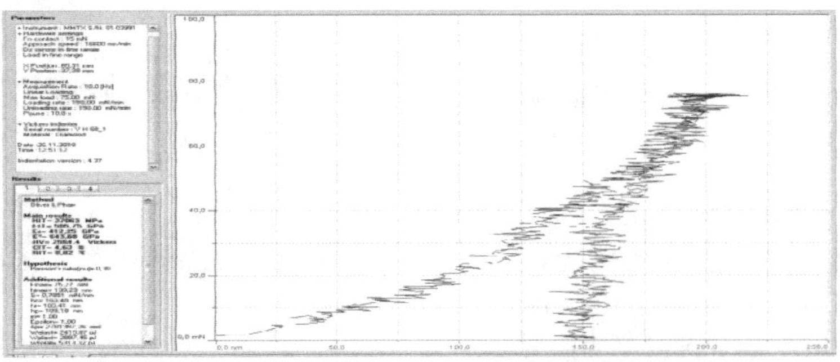

c – Coating 4 TiN / NbN.

Figure 2. Photographs of coating TiN / NbN after tests.

The roller screw mechanism (RSM) efficiency is calculated using the formula below [3]:

$$\eta_{fm} \approx \frac{1}{1 + A_{sl.fr}/(A_u + A_{roll.fr.})/(A_u + A_{axial\ fr.})/A_u} \quad (1)$$

where A_u - useful work per turn of the screw, $A_u = FS$;

$A_{sl.fr.}$ - work of sliding friction forces at non-pole coupling of threads per turn of the screw, at half the profile angle $\alpha = \pi/4$.

Useful work and the work of the friction forces can be found by the following proportion:

$$\frac{A_{sl.fr.}}{A_u} = \frac{2f}{1-f}, \quad (2)$$

where f - coefficient of sliding friction separately for each coating;

$A_{roll.fr.}$ - losses on friction of the paired thread in the transfer:

$$\frac{A_{roll.fr.}}{A_u} = \mu \frac{2\pi}{s}(k+1)tg\alpha, \quad (3)$$

where μ - rolling friction coefficient (shoulder of the rolling friction moment);

$A_{axial\ fr.}$ - friction loss of rolling bearings in the transmission.

The calculations involved both non-coated surfaces and clean steel surfaces. For the purity of the experiment, an additional variable was introduced in the form of plastic lubricants and the calculations measured the operation of the transmission both with and without it [4] (table 2).

Table 2. Dry friction coefficients of uncoated steel surfaces.

Steel	Value of the f coefficient
Structural steel	0.56
Chrome steel	0.47
Tool steel	0.65
Stainless steel	0.45
Titanium alloy	0.58

The values of the axial force and the thread stroke in the considered roller-screw transmission will be equal, respectively:

- F = 10 000 N – axial force;
- S = 4 – the course of the thread.

Based on this, the value of useful work will be equal to: A_u= 40 000 J = 40 kJ.

Using the proportion, we find the value of the work of the friction forces:

$$\frac{x}{40} = \frac{2 \cdot 0.3}{1 - 0.3}; \quad x = 34. \quad (4)$$

Thus we get $A_{sl.fr.}$ = 34 kJ.

To determine the rolling friction in the mechanism [5], we need to know:

- Gear ratio k = 3;
- Angle of inclination of the thread turns α= 45°;

Having the values of all the variables of the equation [6], we get the value of the rolling friction work, also taken for convenience as x:

$$\frac{x}{40} = 0.003 \cdot \frac{2 \cdot 3.14}{4}(3+1) \cdot tg\,45^\circ; \quad (5)$$

$A_{roll.fr.} = 1{,}21\ kJ.$

Now find the rolling friction [7] in the supports:

$$\frac{x}{40} = \frac{1.21}{40}; \quad (6)$$
$$X = 1.21$$

Now we have $A_{axial\ fr.}$=1.21 kJ.

At this stage it is already possible to calculate the efficiency [8]:

$$\eta_{fm} = \frac{1}{1 + 34/(40 + 1.21)/(40 + 1.21)/40} \approx 0.54 \quad (7)$$

We get the value of the efficiency in the mechanism under dry and wet friction for structural steel in table 4 [9].

Table 3. The values of efficiency of steels under dry and wet friction.

Dry friction		Wet friction	
Steel	Value of efficiency	Steel	Coefficient of friction
Structural steel	0.37	Structural steel	0.59
Chrome steel	0.48	Chrome steel	0.5
Tool steel	0.25	Tool steel	0.6
Stainless steel	0.31	Stainless steel	0.48
Titanium alloy	0.2	Titanium alloy	0.55

After performing similar calculations using the obtained data, we have the following values in table 5:

Table 4. The values of efficiency of steels under wet friction.

Steel	Value of efficiency
Structural steel	0.52
Chrome steel	0.42
Tool steel	0.32
Stainless steel	0.36
Titanium alloy	0.25

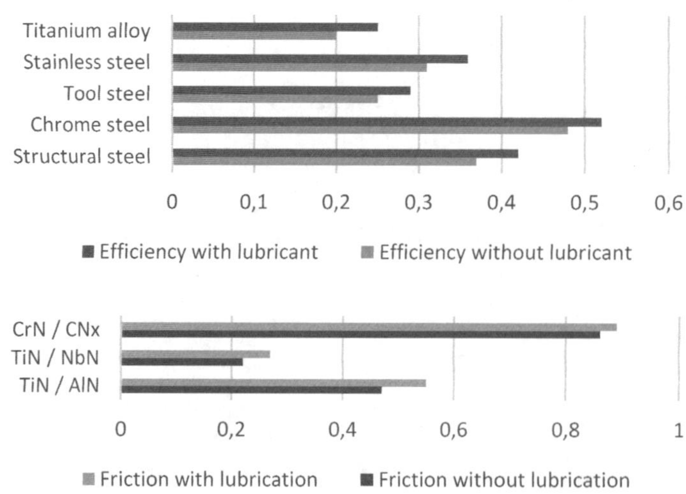

Figure 3. Efficiency of coatings and steels with and without lubricant.

In the case of coatings, the same principles were used in finding the coefficients of friction as with steels, so for the sake of brevity, all the found values of coefficients and efficiency were entered in one General table and two graphs illustrated in figure 3. [10] It reflects the values of the friction coefficients for dry and liquid interaction of parts and the efficiency values corresponding to these coefficients in table 7.

The results of the conducted studies have shown that the use of nanostructured coatings in the operation of mechanisms affects a significant increase in the efficiency of these mechanisms. In particular, the CrN coating shows the best friction characteristics. A slightly less effective coating was based on titanium nitride. In the course of research, it was also found that the most effective steel for these purposes will be structural and chrome steel.

3. Conclusions

Improving the efficiency of mechanical gears due to nanostructured coatings is a poorly studied area, which allows for multiple experiments with both the composition of the applied substances and the materials of the working bodies of mechanisms. Such studies help to study this issue in more detail and form a general understanding of how a particular substance will behave in the workloads of mechanisms.

Acknowledgments

The work was carried out with the financial support of the state task of the Ministry of Science and Higher Education of the Russian Federation (agreement 075-03-2020-046/1, 17.03.2020, subject FZUN-2020-0015, state task of the VlSU GB-1187/20.

References

[1] Hebda M and Chichinadze V 1990 Handbook on tribology **1-3**

[2] Kragelsky I V and Mikhin N M 1983 The friction units of machines *Reference book* p 320

[3] Kragelskii I V and Alisina I V 1978 Friction, wear and lubrication *Handbook* **2** p 400

[4] Berkovich I I and Gromakovsky D G 2000 Tribology. Physical foundations, mechanics and technical applications *Textbook for universities* p 268

[5] Bagmutov V P, Savkin A N and Scab S N 2011 Wear parts of friction units of ground vehicles *The textbook* p 56

[6] Morozov V V, Kochetov D O, Dovbysh N S and Zhdanov A V 2019 Tribological properties of nanostructured coatings for roller screw mechanisms *J. of Phys. Conf. Series* **1331**

[7] Morozov V V, Kosterin A B and Zhdanov A V 2018 Efficiency of Roller–Screw Mechanisms *Rus. Engineer Res.* **38** pp 263-267

[8] Bonanno A, Raimondo M and Pinelli M 2019 Use of Nanostructured Coating to Improve Heat Exchanger Efficiency *Factories of the Future* pp 275-292

[9] Bonanno A, Raimondo M and Zapperi S 2019 Surface nano-structured coating for improved performance of axial piston pumps *Factories of the future* pp 295-314

[10] Pastorello L, Bonanno A 2015 Application of Nano-structured coatings to heat transfer surface of heat exchangers *The Fourteenth Scandinavian International Conference on Fluid Power, May 20-22, 2015, Tampere, Finland*

[11] Malavasi Veronesi F, Caldarelli A *et al* 2016 Is a knowledge of surface topology and contact angles enough to define the drop impact outcome? *Langmuir* **6** pp 13

IOP Conf. Series: Materials Science and Engineering 1118 (2021) 012022 doi:10.1088/1757-899X/1118/1/012022

A device for automatic feeding of coated electrodes into a weld pool

A S Gordynets[1] and A S Nepomnyashiy[1,2]

[1]Tomsk Polytechnic University, Tomsk, Russia
[2]Institute of Strength Physics and Materials Science SB RAS, Tomsk, Russia

asgord@tpu.ru, asn39@tpu.ru

Abstract. The paper presents the specification of a device for automatic feeding of coated electrodes into a weld pool and the results of its application. It eliminates subjective assessment methods and professional skills of welders, which contributes to a more stable metal transfer into the weld pool. The research results have shown that automatic feeding of coated electrodes into a weld pool enables to reasonably assess their welding and technological properties. Any devices similar to the developed one and the results of their tests have not been found.

1. Introduction

Shielded metal arc welding (SMAW) is the most versatile way to obtain permanent joints of metal structures for various applications [1]. Until now, experimental joints have been welded manually without using any electrode feeding mechanisms [2]. The quality of the welded joints depends on many factors, but the qualifications of welders have a particularly strong effect. In order to obtain objective, reliable and repeatable results of conducted investigations in the field of SMAW, it is necessary to reduce the influence of the human factor as much as possible. This is especially true for testing equipment and consumables used for SMAW, where this effect is decisive [3, 5].

To exclude the influence of the professional skills of welders on the research results, various devices are typically used that enable to obtain welded joints without human participation. In the 1930s, a welding method with an inclined coated electrode has been invented in the USSR [7]. In this case, the electrode is supported mechanically and lowered into the weld pool under the action of gravity without human participation (Figure 1). Disadvantages of this method (and similar ones) include the impossibility of controlling the required arc length, as well as constant changes in both the electrode inclination angle and the weld pool coordinates on the horizontal plane. As a result, there is no way to study the kinetics of the electrode melting process because it is difficult to tune focusing of a video camera synchronously with the electrode movement.

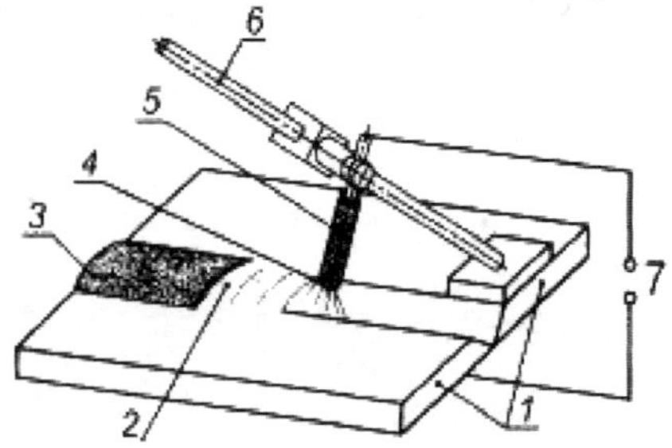

Figure 1. A scheme of SMAW with an inclined coated electrode: 1 – work-pieces, 2 – weld, 3 – slag, 4 – electric arc, 5 – coated electrode, 6 – guide, 7 – power source.

Content from this work may be used under the terms of the Creative Commons Attribution 3.0 licence. Any further distribution of this work must maintain attribution to the author(s) and the title of the work, journal citation and DOI.
Published under licence by IOP Publishing Ltd

2. The device description

To solve this issue, a device for automatic feeding of coated electrodes has been developed, which enables to strike an arc by contacting with work-pieces, to control and stabilize the average arc length upon the electrode melting, as well as to turn it off by the arc elongation. A feature of the developed device is the electrode movement along its axis. As a result, the spatial electrode tip coordinates do not change during the welding process, which enables its video filming [4, 6]. The device (Figure 2) includes the following components: an electrode holder (a); a hull (b), a guide (c), a servo (d), a moving screw (e), and a control unit (does not shown in the figure).

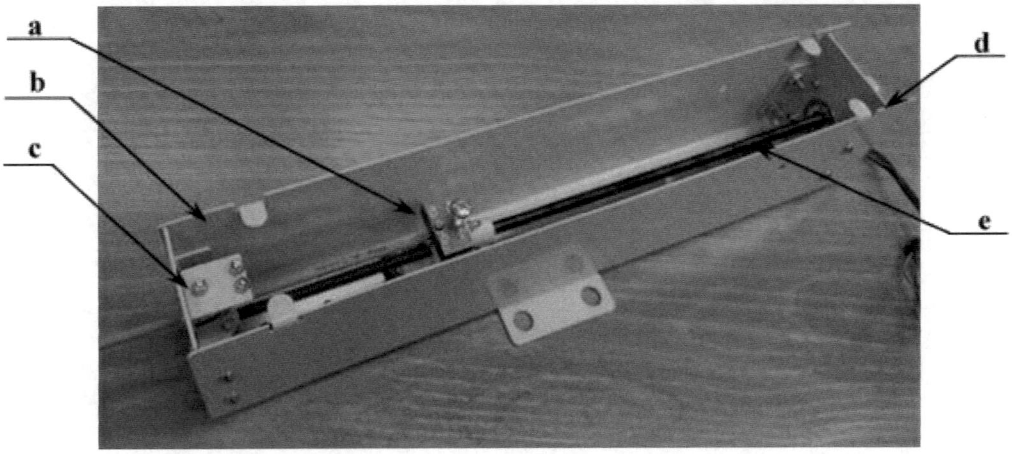

Figure 2. The device for automatic feeding of coated electrodes.

The control unit recognizes its states (no-load, arcing, or short circuit) by measuring the voltage drop in the gap between an electrode and work-pieces and, in accordance with the operator's commands, controls the servo motor. The average arc length is stabilized by comparing the preset value with the actual one. The resulting difference is transmitted to input of a proportional-integral-derivative (PID) controller, which output signal determines both speed and direction of the servo motor rotation.

In order to identify advantages and drawbacks of the developed device, experimental studies have been carried out. Their results have been compared with the data obtained for similar conditions with manually feeding of coated electrodes into a weld pool.

3. Experimental technique, materials and equipment

A scheme of the experiment is shown in Figure 3a. Surfacing of horizontal beads on the 0.12% C – 1.5% Mn – 0.6% Si steel plate with a thickness of 10 mm has been carried out with the OK46.00 and OK74.70 coated electrodes 4 mm in diameter. Welding current (DCEP) has been supplied with a 'VD-306E' conventional three-phase welding rectifier. The θ inclination angle between the electrodes and the plate has been 70°, the arc length has been maintained as low as possible for the both electrode feeding methods. The $V_{welding}$ welding speed has been maintained at 180 mm/min using a special mechanism that has moved the plate. After the arc striking and the surfacing process stabilization, welding current and arc voltage have been recorded for 10 s using a research setup [4]. Obtained oscillograms (Figure 3b) have been analyzed according to the criteria characterizing the welding process stability. The results of the analysis for the OK46.00 and OK74.70 electrodes are presented in Tables 1 and 2, respectively, where U_{rms} is rms arc voltage, I_{rms} is rms welding current, N_{sc} is a number of short circuits during the registration period, t_{mean} is average duration of the short circuits, t_{rsd} is the variation coefficient for the short-circuit durations, T_{mean} is average duration between short circuits, and T_{rsd} is the variation coefficient for the durations between the short circuits.

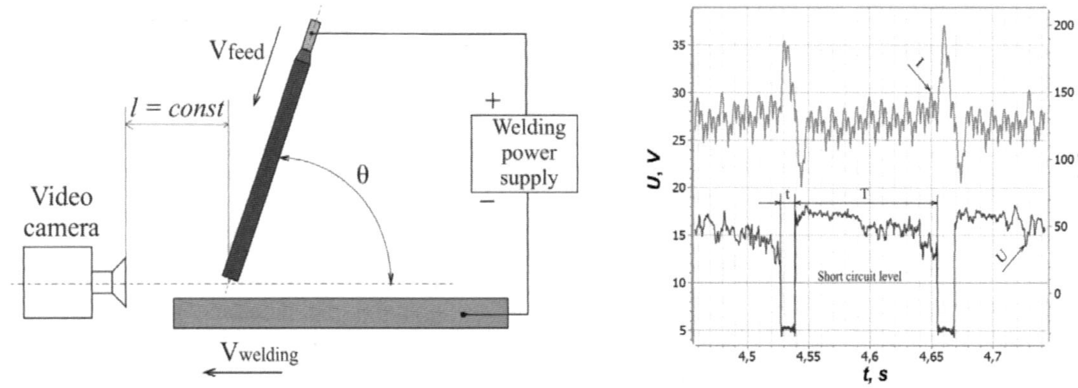

a) **b)**

Figure 3. The scheme of the experiment, as well as welding current and arc voltage oscillograms obtained upon SMAW with automatic feeding of the OK46.00 electrode.

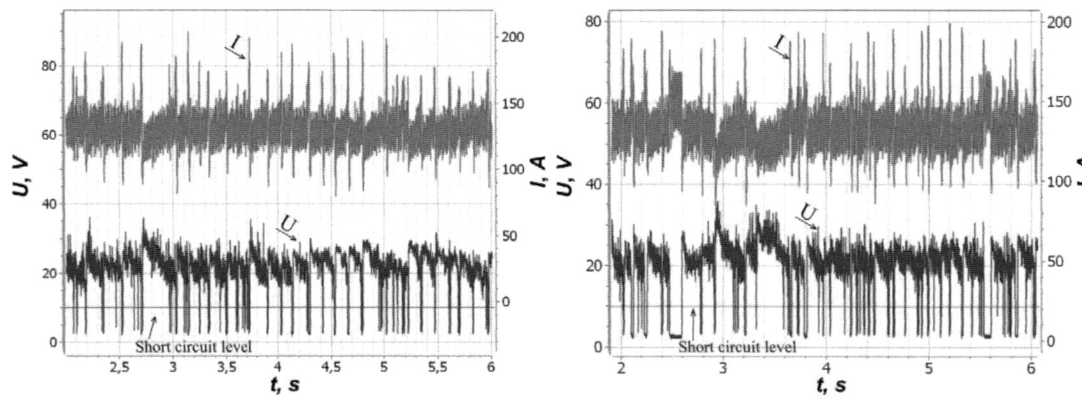

a) **b)**

Figure 4. Current and arc voltage oscillograms in various methods of supplying coated electrodes (a – automatic, b – manual).

Table 1. The parameters of the surfacing process stability with the OK46.00 electrodes.

Electrode feeding methods	U_{rms}, V	I_{rms}, A	N_{sc}	t_{mean}, ms	t_{rsd}	T_{mean}, ms	T_{rsd}
Automatic	21.4	132	74	7.4	0.58	123	0.32
Manual	19.7	134	95	8.9	1.30	121	0.42

Table 2. The parameters of the surfacing process stability with the OK74.70 electrodes.

Electrode feeding methods	U_{rms}, V	I_{rms}, A	N_{sc}	t_{mean}, ms	t_{rsd}	T_{mean}, ms	T_{rsd}
Automatic	22.8	129	26	13.6	0.22	313	0.22
Manual	21.8	139	29	14.9	0.31	360	0.35

The results of the experimental data analysis have showed that the use of the developed device increases the stability parameters of the surfacing process with the coated electrodes. In particular, the deviation from the average values has been significantly reduced while maintaining the values of the average duration between short circuits and their frequency (figure 4). Thus, the use of automatic electrode feeding has caused a decrease in both the variation coefficient for the durations between the short circuits and their frequency by 1.5 times for the OK74.70 basic coated electrodes. Automatic feeding of the OK46.00 rutile-cellulose coated electrodes into the weld pool has reduced the variation coefficient for the durations between the short circuits by 2.24 times, and their frequency by 1.3 times. In this case, both the welding current and arc voltage values have been almost unchanged.

4. Conclusions

The use of the developed device for automatic feeding of coated electrodes (instead of a welder) results in the increase in the stability of the metal transfer to the weld pool, while it gives almost no effect on the energy performance of the welding process. It can be assumed that applying the device improves the mechanical properties of the weld metal. However, more research is required to verify this assumption.

Acknowledgement

Studies of the surfacing processes have been carried out using the research complex created at the expense of the Russian Science Foundation, project 16-19-10010 P.

References

[1] V V Frolov 1988 Theory of welding processes *Moscow: Higher school* 559 p

[2] V S Milyutin and R F Kataev 2016 Welding properties of arc welding equipment *Moscow: NAKS Media LLC* 464 p

[3] Y N Saraev, D A Chinakhov, D I Ilyashchenko, A S Kiselev and A S Gordynets 2017 Investigation of the stability of melting and electrode metal transfer in consumable electrode arc welding using power sources with different dynamic characteristics *Welding International* 31(10) 784–790 doi: 10.1080/09507116.2017.1343977

[4] Y N Saraev, A G Lunev, V M Semenchuk and A S Nepomnyashchii 2020 Heat and mass transfer kinetics in arc welding process *Russian Physics Journal* 62(9) 1573–1579 doi: 10.1007/s11182-020-01878-y

[5] V S Milyutin, R F Kataev and A V Polykhin 2014 Estimation of mode stability when testing power supplies for manual arc welding *Welding and diagnostics* 3 32-38

[6] Y N Saraev, V P Bezborodov, Y V Selivanov and I V Nikonova 2008 Influence of surfacing modes on corrosion resistance of austenite steel welded joints in acidic environments *Assembly in mechanical engineering, instrument making* 5 32–35

[7] A N Kornienko 2004 Welding history. XV – mid XX centuries *Kiev: Phoenix* 212 p

Results of modeling the operation of the hydraulic pump in the Matlab-Simulink application and testing on the diagnostic bench

Yu Vlasov[1], M Meshcheriakov[1] and T Alushkin[2]

[1]Tomsk State University of Architecture and Civil Engineering, 2, Solyanaya Sq., Tomsk, Russia, 634003

[2]Tomsk agricultural Institute branch of the Novosibirsk agricultural University, 19, Karl Marx St., Tomsk, Russia, 634009

E-mail: yury2006@yandex.ru

Abstract. The article provides information on the design of the original stand for diagnostic tests of hydraulic equipment and lubrication systems of various transport and self-propelled vehicles. The scientific novelty of the work is the concept of a mobile diagnostic stand with extended testing capabilities at various drive shaft speeds. As an example of the functional capabilities of the stand, the test results of the NSh-32 pump of the fourth class of the company "Hydrosila" are presented. No publicly available data on the characteristics of the investigated hydraulic pump was found. Based on the test results, a comparative analysis and discussion of the performance of this sample with a pump of the previous class was carried out. Additionally, data on the capabilities of the measuring complex for measuring the parameters of an internal combustion engine, in which the hydraulic pump under consideration can operate, is presented.

1. Introduction

The research aimed at improving the technical service system for transport and self-propelled vehicles through the creation of diagnostic equipment is relevant. More and more complex types of mathematical analysis are being added to the existing diagnostics system [1] and methods are being introduced that were not previously used for hydraulic systems [2]. There is a serial test equipment that allows you to check the performance of hydraulic and lubrication systems of units transmitting torque [3, 4, 5]. As a rule, when diagnosing these systems, a comprehensive assessment of their performance indicators is not carried out [6, 7], imitation of the load on the unit is not provided [5], in addition, such equipment is stationary, uses an electric drive. A three-phase electrical network is required for operation. To solve the above problems, it is proposed to use a unit with a hydraulic system as diagnostic equipment. General view and hydraulic diagram of the unit are shown in Figure 1. Principle of unit operation. The hydraulic oil from the tank (1) is cleaned in a full-flow filter (3), then enters the hydraulic pump (9) mounted on the tested unit. By changing the fuel supply of the internal combustion engine and the position of the adjustable throttle (7), the characteristic speed and load modes of operation of the tested unit are established. The rotation speed is controlled by standard or portable tachometers, the oil pressure is measured by a pressure gauge (6) on the unit dashboard. After setting the characteristic operating mode by switching the on / off valve control knob (8), the oil

Content from this work may be used under the terms of the Creative Commons Attribution 3.0 licence. Any further distribution of this work must maintain attribution to the author(s) and the title of the work, journal citation and DOI.
Published under licence by IOP Publishing Ltd

flow is directed to the positive displacement flow meter (4), the readings of which are necessary for calculating the load parameters. After passing through the flow meter, the oil returns to the unit tank. To prevent pipe leaks, a block with an adjustable safety valve (5) is installed in front of the adjustable choke. During the tests, the following operating parameters are monitored:
- pressure in the high-pressure line;
– oil temperature in the hydraulic tank;
– oil consumption over time;
– shaft rotation frequency (option). The use of the developed unit with a hydraulic drive makes it possible to realize the following advantages:
– smaller weight and size parameters;
– mobility, since the developed installation is located on the chassis;
– ensuring smooth control of the rotation speed;
– the ability to check both engines and gearboxes in connection with the relatively low speeds of the hydraulic pump, as well as diagnose the parameters of the operation of hydraulic pumps at different modes of shaft rotation speed.

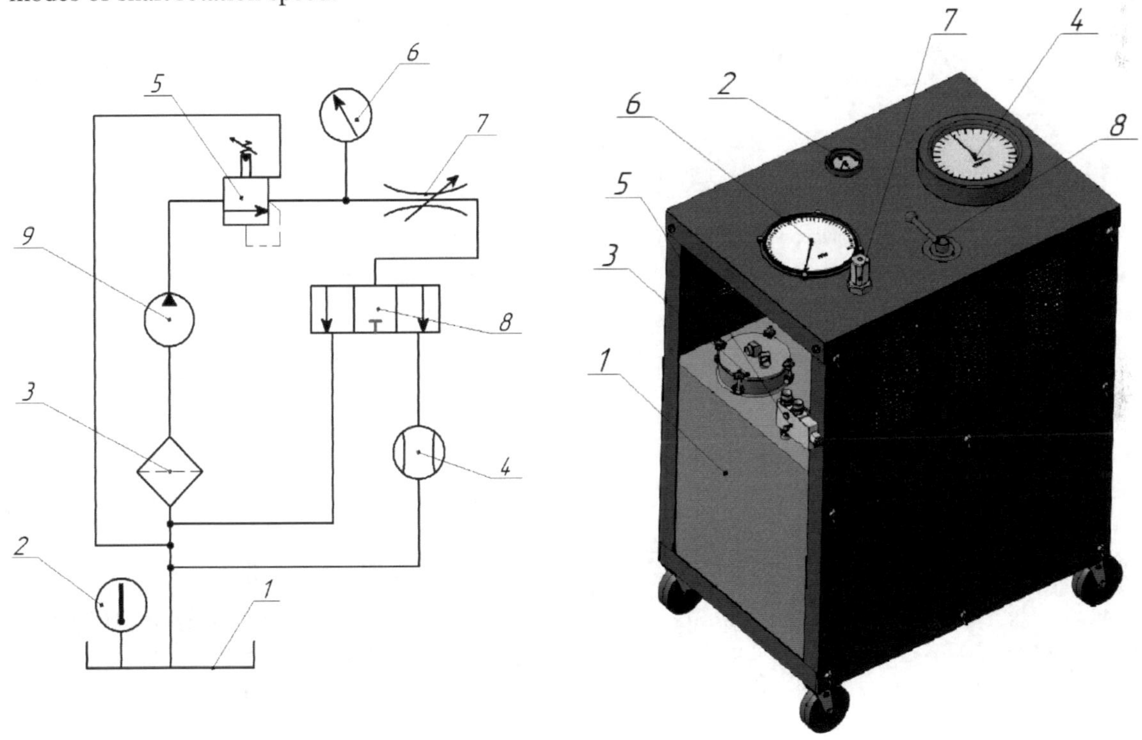

a) *b)*

Figure 1. Scheme of the stand:
a) hydraulic scheme of the stand; b) model view illustration.

2. Materials and Methods

The hydraulic pump NSh-32 manufactured by the company "Gidrosila" [4] was selected as the object of diagnostics. The choice of the model is due to its widespread use in transport and self-propelled vehicles. Due to the lack of data on the parameters of the fourth generation pumps, a numerical mathematical modeling of the operation of the NSh-32 pump of the MASTER series of the fourth generation was carried out. The simulation of the operation of the hydraulic pump was carried out using the engineering calculation program of the Matlab-Simulink application (Figure 2). To check the adequacy of the simulation on the designed stand, a series of tests of the specified pump was carried out and the values obtained were evaluated with the parameters of the NSh-32M pump of the third generation (Figure 3). The following constants were used in the simulation and testing: the kinematic viscosity of the VMGZ hydraulic oil $0.75 \cdot 10^{-4}$ m²/s, the mechanical efficiency of the pump 0.95. The rotational speed of the drive shaft is 25 s^{-1} (1500 rpm).

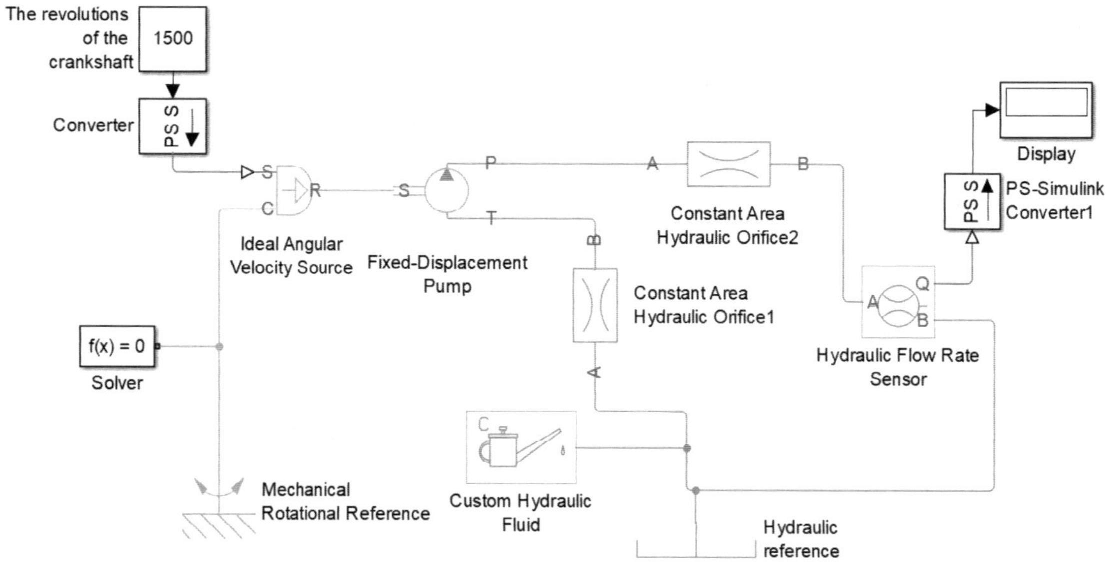

Figure 2. Designed circuit in the Simulink production environment.

3. Results

Based on the results of numerical modeling in the Matlab - Simulink application, a dynamic change in various physical indicators, for example, the pressure and flow of hydraulic oil, at selected points of the designed circuit was obtained. Based on the test results of the selected pump brand on the bench, the characteristic was obtained in the steady-state operation mode (Figure 3). It was found that for a fourth-class hydraulic pump the maximum value of the volumetric efficiency is 0.86 and is achieved in the range of 12.5 ... 14 MPa. For comparison, the same pump model of the third class achieves a volumetric efficiency of 0.88 at a pressure in the range of 10 ... 12 MPa.

4. Discussion

The simulated work processes of the designed stand in the Matlab-Simulink application cannot be evaluated based on the test results of the hydraulic pump on the proposed stand, since the installed measuring devices do not allow evaluating the dynamic change in the recorded values. However, the data obtained during the tests make it possible to evaluate the performance of hydraulic pumps at a steady-state pump shaft speed, as well as to additionally carry out tests at other speed modes, if necessary.

In the near future, it is planned to conduct research on the operating modes of the NSh-32 hydraulic pump installed on the D-144 engine. Research will include the measurement of pressure parameters, volumetric flow rate of hydraulic fluid at a given engine speed. The use of an internal combustion engine as a drive for a hydraulic pump will make it possible to investigate the indicated parameters at wider ranges of rotation frequencies. It will also be possible to evaluate engine performance such as power, torque, hourly fuel consumption and specific effective fuel consumption.

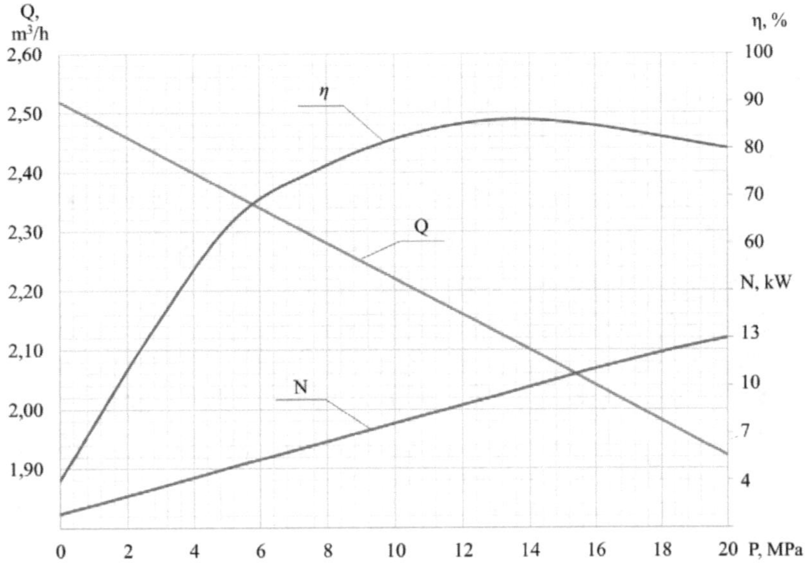

Figure 3. The obtained characteristics of the NSh-32 of the fourth class.

5. Conclusion

The simulation showed the limited capabilities of the application used. In subsequent studies of the hydraulic pump, the rotational speeds of the drive shaft will be varied in order to build a multi-parameter characteristic. Additionally, the use of a pump as a loading element will allow simulating the load on the units in order to diagnose the parameters of the lubrication system.

At the moment, an experimental measuring complex installed on the engine can measure the pressure and temperature of the oil in the engine lubrication system. If necessary, it is possible to realize the measurement of the volumetric flow rate of the oil pump.

References

[1] Przystupa K, Ambrożkiewicz B, Litak G 2020 Diagnostics of transient states in hydraulic pump system with Short Time Fourier Transform Advances in Science and Technology – Research Journal 14(1) pp 178-183

[2] Liu Y, Yao E, Xu H 2012 Fault diagnosis of hydraulic pump based on particle filtering and autoregressive spectrum. Chinese Journal of Scientific Instrument, 33(3) pp 561-567

[3] Nosihin A 2012 Modular diesel engine run-in stand CyberLeninka "Mechanics and Mechanical Engineering" 4 pp 59-62

[4] Petrishchev N, Sayapin S, Dankov A and Kapustkin A 2013 Universal control and diagnostic service equipment for the maintenance and repair of hydraulic drive units, transmissions of the machine and tractor fleet *Construction Mechanization Magazine* 1 (823) pp 5-10

[5] Kapustkin A 2013 Diagnostics of the technical condition of centrifugal filters for oil purification at the stand KI-28256.01 *Proceedings of GOSNITI* 1 pp 13-17

[6] Fedotova M, Kovalev A, Kushel A et al. 2017 *Evaluation of machines and equipment* Moscow: INFRA-M

[7] Barylnikova E, Gafiyatullin A and Kulakov O 2015 Diagnostics of wear rate of diesel engine bearings during operation *Problems of the functioning of transport systems. Materials of the international scientific-practical conference of students, graduate students and young scientists* Tyumen: Tyumen Industrial University pp 68-75

Thermal conductivity and heat capacity of nanofluid based on water modified by hybrid material of composition detonation nanodiamonds-carbon nanotubes

A A Vozniakovskii[1], A P Voznyakovskii[2], S V Kidalov[1] and E K Kalashnikova[1]

[1]Laboratory “Physics for Cluster Structures”, Ioffe institute, Saint-Petersburg, Polytechnic st., 28, 194021, Russian Federation
[2]Institute of Synthetic Rubber, Saint-Petersburg, Gapsalskaya st., 1, 198035, Russian Federation

E-mail: alexey_inform@mail.ru

Abstract. This paper is devoted to assessing the prospects of using a DND-CNT hybrid material to improve water's thermophysical properties compared with pure DND and CNT. A hybrid powdered nanomaterial composed of detonation nanodiamonds-multi-walled carbon nanotubes was synthesized by growing multi-walled carbon nanotubes on the surface of nanodiamonds aggregates coated with catalyst using CCVD method for nanotubes growth. The mass fraction of CNTs in the hybrid material was 30%. Using 0.2 mass. % of the hybrid material, we managed to increase the thermal conductivity of water by 94% and increase the specific heat by 17% (at a temperature of 60 degrees) without changing the kinetic viscosity of water. Using initial detonation nanodiamonds or commercial single-walled nanotubes as an additive at the same concentration and temperature, the maximum increase in thermal conductivity was only 37%, and the maximum increase in specific heat was only 5%. Simultaneously, the mechanical mixture of detonation nanodiamonds and carbon nanotubes (in a 7 to 3 ratio) did not form stable water suspensions.

1. Introduction

Cooling liquids are still actively used as heat carriers in cooling systems for various devices, from internal combustion engines to cooling systems for computers' electronic systems. Up to now, one of the most actively used cooling liquid is water. Water received such distribution as a cooling liquid due to its characteristics: the thermal conductivity of water at 20°C is 0.6 W/(m*K); the specific heat is 4180 J/(kg*°C), while the dynamic viscosity is 1004 mPa*s. However, with the development of technology, the amount of heat that must be removed using the cooling liquid is constantly increasing, which leads to the impossibility of efficient heat removal using water without increasing the volume of the water used. Besides, due to the relatively low thermal conductivity of water, it is necessary to increase the radiators' area for effective heat dissipation, which is also not always possible. Therefore, many scientific groups are working on creating new coolants based on water to improve their thermophysical properties. Carbon materials (CNM), such as detonation nanodiamonds (DND), carbon nanotubes (CNT), and graphene nanostructures, are actively used as modifying additives. Stable CNM suspensions in various liquid media are called nanofluids (NF). The choice of CNM as additives for the creation of NF is due to their high thermophysical properties. For example, the DND thermal conductivity can be

Content from this work may be used under the terms of the Creative Commons Attribution 3.0 licence. Any further distribution of this work must maintain attribution to the author(s) and the title of the work, journal citation and DOI.
Published under licence by IOP Publishing Ltd

estimated as typical for single-crystal diamond (type IIa) 2000 W/(m*K) at room temperature [1], the thermal conductivity of single-walled CNTs (SWCNTs) along the tube is estimated at 3000 W/(m*K) [2], and the thermal conductivity of single-layer graphene is 5000 W/(m*K) [3]. Researchers are also trying to obtain new hybrid materials of DND-CNT composition and graphene nanostructures-CNT. The motivation for the synthesis and use of hybrid materials is the disappearance of the disadvantages inherent in one of the materials (for example, the complexity of stabilizing CNTs in water without the use of surfactants) and the manifestation of a synergistic effect [4].

The purpose of this study was to assess the prospects of using a DND-CNT hybrid material to improve the thermophysical properties of water in comparison with pure DND and CNT.

2. Experimental section

2.1. Materials
As the initial materials for the creation of water-based nanofluids, we used commercial DND with a particle size of 4-5 nm (manufactured by SKTB "Technolog", Russia) [5], single-walled carbon nanotubes (manufactured by "OCSiAl" company, Russia), DND-CNT hybrid material and filtered water (the electrical conductivity of water was 270 μS/cm). A hybrid powdered nanomaterial composed of detonation nanodiamonds-multi-walled carbon nanotubes was obtained by growing multi-walled carbon nanotubes on the surface nanodiamonds aggregates coated with catalyst using catalyst chemical vapor deposition (CCVD) method for nanotubes growth. The mass fraction of CNTs in the hybrid material was 30%. The procedure for the synthesis of a hybrid material is described in detail in [6].

2.2. Scanning electron microscopy (SEM) study
The study of the DND-CNT hybrid material was performed on SEM microscope Supra55VP-3259 (Zeiss, Germany).

2.3. Synthesis of nanofluids
For the synthesis of nanofluids, powders of carbon nanomaterials (DND; SWCNT; powder mixture of DND and SWCNT in a ratio of 7 to 3 and DND-CNT hybrid material) were added to a beaker with water with constant stirring. The beaker was then placed in an ultrasonic bath where the sample was sonicated for 30 minutes with a pause of 5 minutes after every 10 minutes of ultrasonic treatment to prevent overheating of the sample. The concentration of carbon materials was 0.2 wt. %.

2.4. Measurement of particle size and zeta potential of DND-CNT hybrid material
The measurement of the particles size and zeta potential was performed with a dynamic light scattering (DLS) method on the device ZetaSizer Nano ZS Malvern (USA). A stable suspension of DND-CNT hybrid material (0.2 wt. %.) was taken for measurement.

2.5. Dynamic viscosity measurement
Dynamic viscosity measurement was performed on the rotational viscometer Brookfield HBDV-E (Germany).

2.6. Measurement of thermophysical properties of nanofluids
The thermophysical properties of nanofluids (thermal conductivity, specific heat) were performed with a laser flash method on the device LFA-457 (NETZSCH, Germany). The error in measuring the thermal conductivity was 5%, and the error in measuring the thermal conductivity was 3%. The temperature during measurements was 60°C.

3. Results and discussion
Figures 1 and 2 show the results of scanning electron microscopy of the DND-CNT hybrid material and measuring the particle size of the DND-CNT hybrid material in water by the DLS method, respectively.

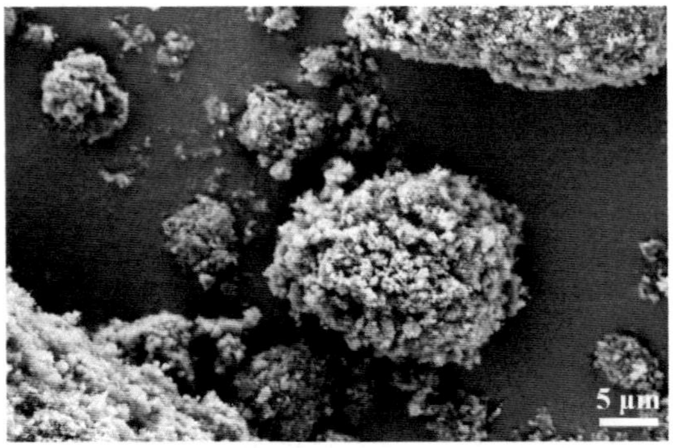

Figure 1. SEM image of DND-CNT hybrid material; scale ruler – 5 µm.

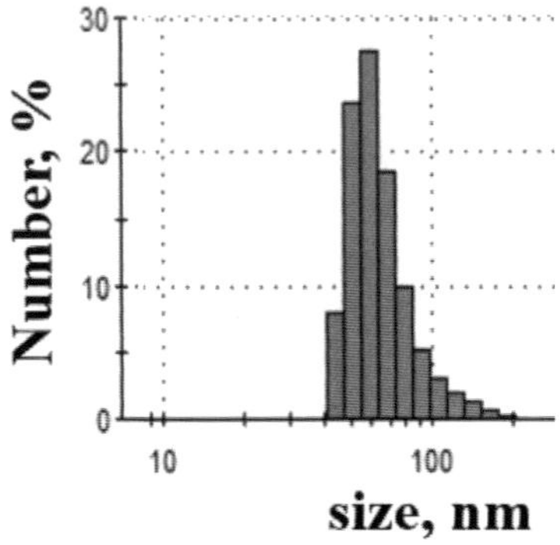

Figure 2. Particle size of DND-CNT hybrid material.

As shown in Figure 1, the hybrid material in the form of a powder is micron-sized aggregates. However, as shown in Figure 2, these easily destructible aggregates are composed of stable agglomerates with an average of 50-60 nm. Based on the data obtained, a model of the hybrid material structure was proposed, shown in Figure 3. To assess the stability of the obtained aqueous suspensions, the samples of suspensions obtained after ultrasonic treatment were placed in plastic cuvettes and photographed 24 hours (Figure 4).

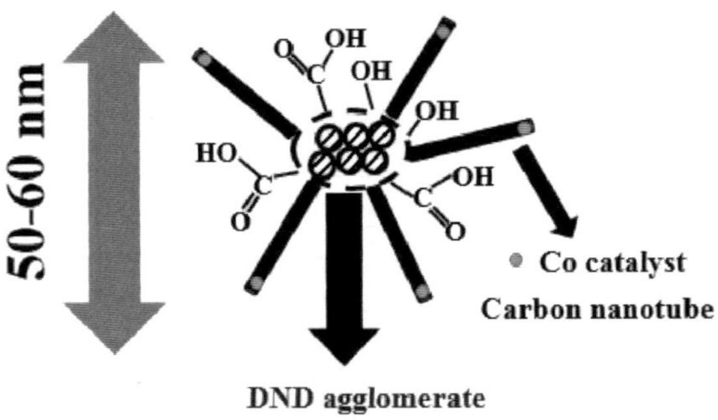

Figure 3. The structure of the DND-CNT hybrid material.

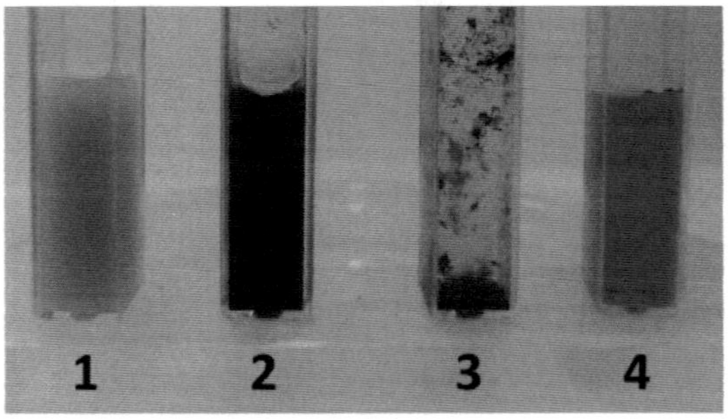

Figure 4. Photos of suspensions of the DND, (1) CNT (2), DND-CNT mixture (3), and the DND-CNT hybrid material (4) after 24 h of settling.

As shown in Figure 3, all materials, except for the mixture of DND and SWCNT, form stable suspensions in water. Therefore, to study the thermophysical properties (thermal conductivity, specific heat capacity), DND, SWCNT, and DND-CNT hybrid material nanofluids in water were taken. The working temperature at which liquid coolants are used in the technology is 50-80 degrees. Therefore, a temperature of 60 degrees was chosen to measure the thermophysical properties. The results of thermal conductivity measurements and the heat capacity of nanofluids are shown in Figures 5 and 6, respectively.

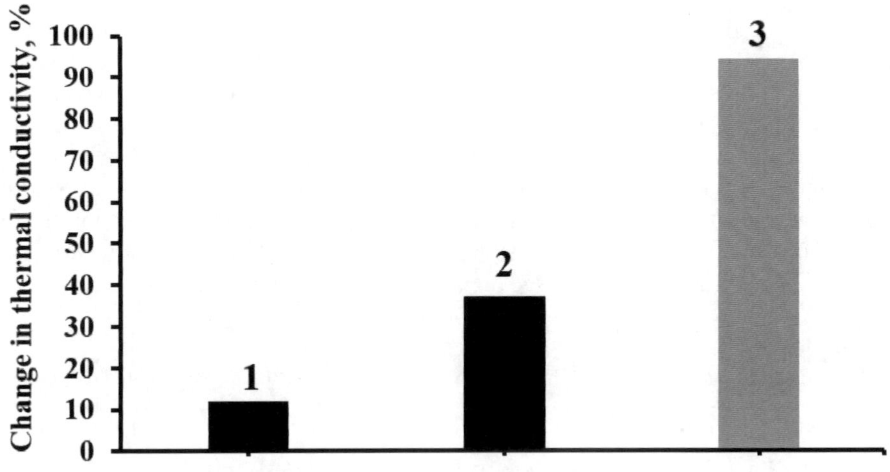

Figure 5. Change in thermal conductivity (%) of aqueous suspensions of DND (1), CNT (2) and DND-CNT hybrid material in comparison with the initial water.

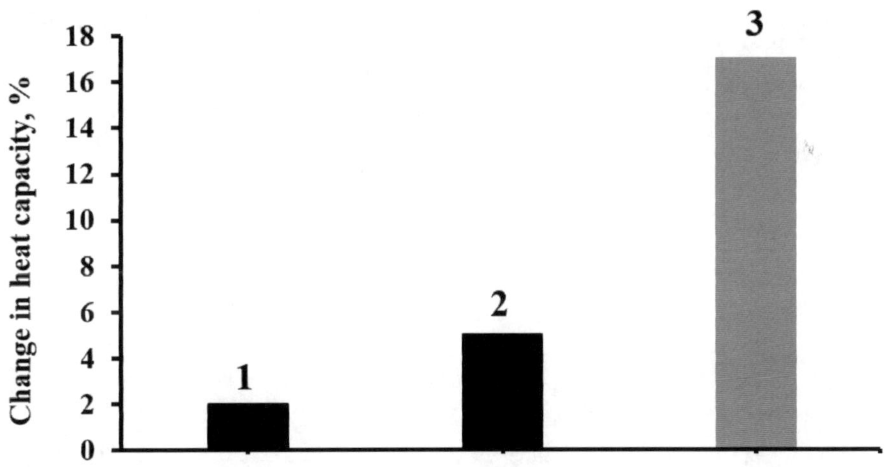

Figure 6. Change in specific heat (%) of aqueous suspensions of DND (1), CNT (2) and DND-CNT hybrid material in comparison with the initial water.

As shown in Figure 5, the use of a hybrid material made it possible to increase thermal conductivity by 94% compared to the original water. Simultaneously, the use of DND and SWCNT made it possible to increase the increase in thermal conductivity only by 12 and 37% compared to the initial water. As shown in Figure 6, the use of a hybrid material made it possible to increase the increase in specific heat capacity by 17% compared to the original water. Simultaneously, the use of DND and SWCNT made it possible to increase the increase in thermal conductivity only by 2 and 5% compared to the initial water.

The DND-CNT hybrid material nanofluid's dynamic viscosity was 0.471 mPa*sec., which is comparable to the initial water's dynamic viscosity at the same temperature (0.467 mPa*sec.).

5. Conclusion

It was found that even though the use of a DND-CNT hybrid material even at low concentrations (0,2 wt. %) allows one to synthesize more efficient cooling nanofluids based on water in terms of thermophysical properties that than using pure DND and SWCNT. Although the CNTs included in the DND-CNT hybrid material are hydrophobic and do not form stable suspensions in water, the hybrid material itself easily forms stable aqueous suspensions without surfactants. It is important to note that the particles of the hybrid material with a size of 50-60 nm can increase the thermal conductivity of base water as well as increase its heat capacity, which is an unexpected result that is important from a practical point of view. The obtained results of the study of thermophysical properties clearly demonstrate the promising nature of hybrid materials for the creation of efficient cooling nanofluids.

Acknowledgments
The reported study was funded by RFBR and BRFBR, project number 20-58-00056.

References

[1] Speight J G 2004 *Lange's Handbook of Chemistry (16th ed.)*; (USA:New York/ McGrawHill) section 4 Table 4.1
[2] Pop E, Mann D, Wang Q, Goodson K and Dai H 2006 *Nano Lett.* **6** 96
[3] Balandin A A, Ghosh S, Bao W, Calizo I, Teweldebrhan D, Miao F and Lau C N 2008 *Nano Lett.* **8(3)** 902.
[4] Aravind S S J, Ramaprabhu S 2013 *RSC Adv.* **3(13)** 4199
[5] Aleksenskiy A E, Eydelman E D, Vul A Y 2011 *Nanosci. Nanotechnol. Lett.* **3(1)** 68
[6] Vozniakovskii A A, Kol'tsova T S, Voznyakovskii A P, Kumskov A L, Kidalov S V 2020 *J. Colloid Interface Sci.* **565** 305

IOP Conf. Series: Materials Science and Engineering 1118 (2021) 012025 doi:10.1088/1757-899X/1118/1/012025

Study of the magnetic property of supermagnetic fluid

O K Kuvandikov, Sh J Kuvandikov, M K Salakhitdinova and Kh A Kayumov

Faculty of Physics, Samarkand State University, Samarkand 140104, Uzbekistan

E-mail: kayumov0130@gmail.com

Abstract. In this work, the magnetic fluids based on Fe_3O_4 have been obtained by the chemical co-precipitation method. Analysis of the size and morphology of the synthesized iron oxide nanoparticles of the magnetic fluid using transmission electron microscopy revealed that nanoparticles are of 25-38 nm in size and spherical in shape. The elemental composition of magnetite nanoparticles of the fluid was examined in energy-dispersive X-ray analysis and determined to contain only Iron and Oxygen. The magnetic property of the magnetic fluid was measured using vibrating sample magnetometer. The results showed that the magnetic fluid has a superparamagnetic character which of specific saturation magnetization of 3.4 emu/g

1. Introduction

Since the 60s of last century, the physical properties of the magnetic fluids based on magnetite nanoparticles are being studied. According to their mechanical, thermal, electrical, magnetic, and other properties, these materials are widely used in various science fields. For example, such liquids have found numerous applications in biomedicine, electronics, industry mining, machinery, hyperthermia, and other fields [1,2]. Magnetic fluids are colloidal dispersed systems consisting of stabilized ferro or ferromagnetic nanoparticles suspended in liquid carriers [3]. The technological process to obtain magnetic fluids consists of two stages [4]. The first one is the synthesis of magnetic nanoparticles. The second is stabilizing of the synthesized nanoparticles in a carrier liquid by surfactants or polymers [5]. As of today, there is a variety of methods to synthesize magnetic nanoparticles, such as co-precipitation, hydrothermal, sol-gel, thermal decomposition, solvothermal, sonochemical, electrochemical, mechanical milling and other methods [6,7]. In the synthesis of the Fe_3O_4, the chemical co-precipitation method was acceptable because of its simplicity, convenience, low cost, and most importantly, the ability to control the size of the particles. The surfactant is used to prevent aggregation of the nanomagnetic particles of magnetic fluid. Generally, surfactant molecules have a polar "head" and a non-polar "tail" (or vice versa). One of the ends is adsorbed to the particle, and the other is attached to the liquid carrier molecules, forming a normal or reverse micelle around the particle, respectively [8]. We used sodium oleate as a surfactant in the preparation of the liquid.

If we consider the width of the field of application of magnetic fluids, the study of their magnetic properties at room temperature is actual. Therefore, this research aims to synthesize the magnetic fluid based on iron oxide nanoparticles and study their morphology, element composition, and magnetization.

2. Materials and methods

In the preparation of the magnetic fluids, Fe_3O_4 nanoparticles were initially synthesized by the chemical co-precipitation method. Each of the salts of $FeCl_3 \cdot 6H_2O$, $FeSO_4 \cdot 7H_2O$ with the molar ratio

Content from this work may be used under the terms of the Creative Commons Attribution 3.0 licence. Any further distribution of this work must maintain attribution to the author(s) and the title of the work, journal citation and DOI.
Published under licence by IOP Publishing Ltd

of Fe^{2+}: Fe^{3+}=1:2 was dissolved separately in 100 ml of double distilled water, then was mixed together at 60 ^{0}C using a magnetic stirrer for 30 min. Afterward,1M KOH as the precipitating agent was slowly added to the resulting salt solution drop by drop until the pH 7 and stirred at 100 ^{0}C 30 min. Then the precipitated nanoparticles were washed with double-distilled water several times to remove the impurities. The sediment was filtered, dried in the oven at 100 ^{0}C for 6 hours and pulverized to a powder. The reaction equation of the experiment is shown below:

$$2FeCl_3 \cdot 6H_2O + FeSO_4 \cdot 7H_2O + 8KOH \Rightarrow Fe_3O_4 \downarrow + K_2SO_4 + 6KCl + 23H_2O \quad (1)$$

Magnetic fluids were prepared under the condition that 1 g of sodium oleate dissolved in 50 ml of distilled water was added to the sediment as a surfactant to prevent particle agglomeration, then mixed well at 80 ^{0}C for 1 hr.

Morphology of the nanoparticles of the magnetic fluids and elemental composition of the Fe_3O_4 powders have been examined using a transmission electron microscope (model: TEM LEO 912 AB) and an energy dispersive X-ray spectroscopy (EDX, model: X-ACT Silicon Drift Detector). The magnetizations of the fluid samples were measured with a vibrating sample magnetometer at the room temperature.

3. Results and discussion

3.1. TEM studies.

The shape, size, and morphology of the nanoparticles of the magnetic fluid were investigated by using transmission electron microscope (TEM). The obtained results are shown in figure 1. From TEM images it can be seen that almost all of the particles of the samples are spherical in shape. We can also see that the size of the particles are in the range from 25 to 38 nm. These results confirm that monodisperse and spherical magnetite nanoparticles were formed.

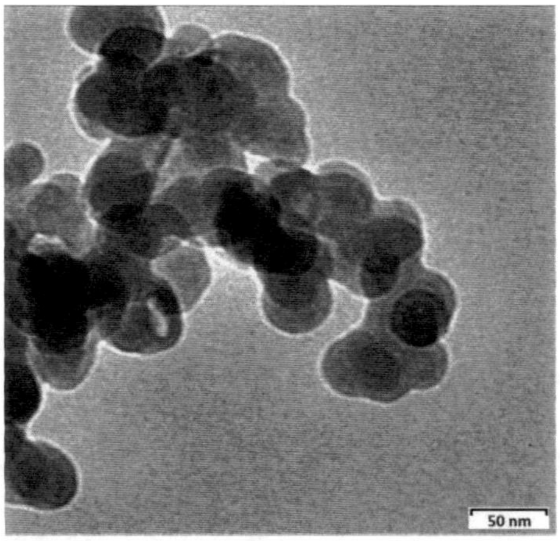

Figure 1. TEM image of the Fe_3O_4 nanoparticles

3.2. Chemical composition studies.

The elemental compositions of the magnetite nanoparticles were analyzed by the energy dispersive X-ray (EDX) measurements. The measurement results are shown in figure 2. As can be seen from figure 2, the sample contains the elements Fe and O, and the value percentage of Fe is higher than O. It is known that magnetite is composed of Fe and O elements. The EDX spectrum indicates that the obtained sample is composed of only Fe and O. However, a small percentage of sulfur is present which might be due to the ferric salts used.

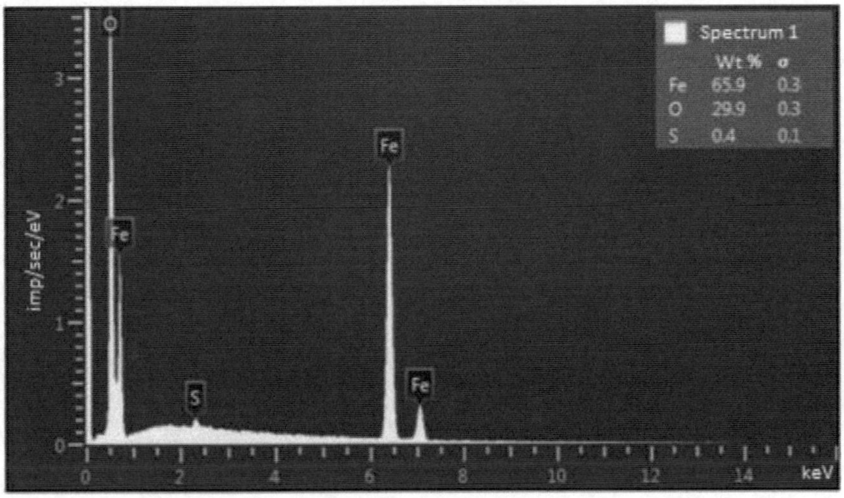

Figure 2. EDX spectra of the magnetite powder

3.3. Magnetic properties.

The dependence of the magnetization of the magnetic fluid on the magnetic field was measured by Vibrating sample magnetometer at room temperature. The results of the measurements are shown in figure 3. From figure 3 it can be noted that the magnetization of the sample increases with increasing magnetic field and attains the saturation. From the magnetization hysteresis loops of the figure 3 we can determine the saturation magnetization, remnant magnetization and the coercivity values. The saturation magnetization value for the Fe_3O_4 magnetic fluid is 3.4 emu/g.

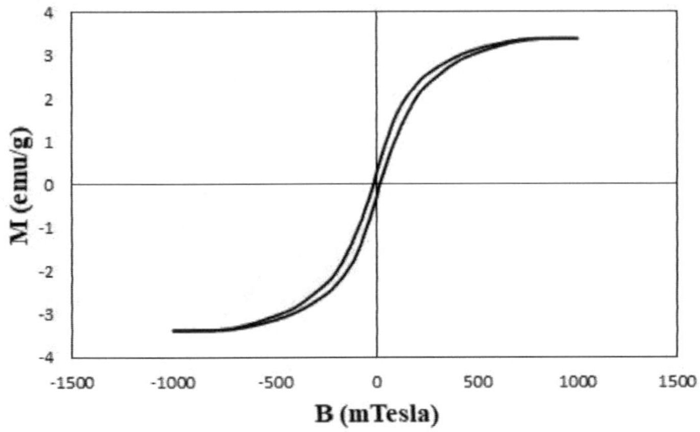

Figure 3. Hysteresis loops of the Fe_3O_4 magnetic fluid

4. Conclusion

Magnetic fluid based on high purity iron oxide nanoparticles have been successfully synthesized using chemical co-precipitation method. Based on EDX analysis it was confirmed that the synthesized particles were Fe_3O_4. TEM measurements revealed that the sizes of the Fe_3O_4 nanoparticles of the magnetic fluid are about 25-38nm. We have also revealed that the magnetite magnetic fluid shows superparamagnetic nature, and its saturation magnetization value was 3.4 emu/g.

References

[1] Sophie Laurent, Silvio Dutz, Urs O Hafeli, Morteza Mahmoudi. 2011 *Advances in Colloid and*

Interface Science Magnetic fluid hyperthermia: Focus on superparamagnetic iron oxide nanoparticles **166** 8–23

[2] Ladislau Vekas, Doina Bica, Mikhail V. Avdeev. 2007 *China Particuology* Magnetic nanoparticles and concentrated magnetic nanofluids: Synthesis, properties and some applications **5** 43–9

[3] Dyachenko S V, Vaseshenkova M A, Martinson K D, Cherepkova I A, and Zhernovoi A I 2016 *Russian Journal of Applied Chemistry* Synthesis and Properties of Magnetic Fluids Produced on the Basis of Magnetite Particles **89, № 5** 690-6

[4] Gorbyk P P, Abramov N V, Petranovskaya A L, Turelyk M P, Pilipchuk E V, Oranskaya E I, Konchits A A, Shevchenko Yu B 2011 *Pverxnost* Sintez i svoystva magnitnix jidkostey na osnove nanorazmernogo Fe_3O_4 **3(18)** 245-7

[5] Scherer C and. Figueiredo A M Neto 2005 *Brazilian Journal of Physics* Ferrofluids: Properties and Applications **35 №3A** 718-27

[6] Kebede K Kefeni, Titus A M Msagati, Bhekie B Mamba 2017 *Materials Science and Engineering B* Ferrite nanoparticles: Synthesis, characterisation and applications in electronic device **215** 37-55

[7] Attarad Ali, Hira Zafar, Muhammad Zia, Ihsan ul Haq, Abdul Rehman Phull, Joham Sarfraz Ali, Altaf Hussain 2016 *Nanotechnology, Science and Applications* Synthesis, characterization, applications, and challenges of iron oxide nanoparticles **9** 49–67

[8] Novopashin S A, Serebryakova M A, and Khmel S Ya 2015 *Thermophysics and Aeromechanics* Methods of magnetic fluid synthesis (review) **22 4** 397-412

Determination of Optimal Frequency to Replace Cutting Elements in Chain Trenchers

S M Kravchenko, V A Slepchenko and I V Slepchenko
Tomsk State University of Architecture and Building, 2, Solyanaya Sq., Tomsk, 634003 Russia

E-mail: kravchenkosm.1951@mail.ru

Abstract. The article deals with some factors influencing the wear resistance of cutting elements in chain trenchers. It provides a number of recommendations to ensure most cost-effective operations with cutting tools in chain trenchers.

1. Introduction

Developing frozen soils is much more laborious than thawing soils [1]. Therefore, mistakes made in the design and operation of cutting elements of trench excavators are very costly. Thus, the urgency of more advanced methods for assessing the state of cutters is obvious.

2. Analytical review

The abrasiveness and strength of frozen soils are an important factor in earthworks. To destroy this environment effectively, it is necessary to carefully monitor the state of cutting elements in chain trenchers, since unacceptable wear leads to a proportional multiple increase in resistance during the operation of a machine [1]. In addition, unacceptable wear on the cutting tools significantly reduces the performance of the trenchers.

Therefore, the main criteria for optimizing the operation of trench excavators are productivity and economic efficiency.

As shown by the studies [1–4], mathematical relationships, though having a common physical meaning for various types of trench excavators, have fundamental differences. Therefore, this work is a continuation of the article [4] adapted for chain trenchers.

3. The concept model of development

The main criterion for making a decision on the service life of cutting elements assembled for chain trenchers is the reduced costs per unit of production [2, 4]:

$$C_{\mathrm{pu}} = \frac{C_{\mathrm{m}} + n_{\mathrm{e}} \cdot K_{\mathrm{c}}}{v_{\mathrm{af}} \cdot k_{\mathrm{t}}} + \frac{C_{\mathrm{re}}}{S}, \tag{1}$$

where C_{m} is the cost of machine-hour taken for a machine operation, excluding replaceable equipment; C_{re} is the cost of replaceable equipment (cutting tools on a chain); S is the length of an excavated trench; n_{e} is the normative coefficient efficiency; k_{t} is the coefficient of transition to operational productivity; v_{af} is the average feed rate during the operation of the cutting assembly; K_{c} is the capital investment for one hour of trencher operation.

As the length of a trench grows, the value of the first component in the equation (1) increases due to an

Content from this work may be used under the terms of the Creative Commons Attribution 3.0 licence. Any further distribution of this work must maintain attribution to the author(s) and the title of the work, journal citation and DOI.
Published under licence by IOP Publishing Ltd

increased wear of the teeth, thus leading to a proportional drop in the productivity of the machine. On the other hand, this process is accompanied with a fall in the costs associated with replacement tooling – the second component in the equation (1). As a result, the function $C_{pu}(v_{af}, S)$ is non-linear. Once applied to frozen soils with sufficient strength and abrasiveness, the dependence (1) has a pronounced extremum (section *BC* in Figure 1), which provides a required reasonable life of the teeth. If we continue to operate the cutting elements without tracking this effect, which takes place in practice, the costs begin to grow significantly (Graph 1 in Figure1) [4, 5].

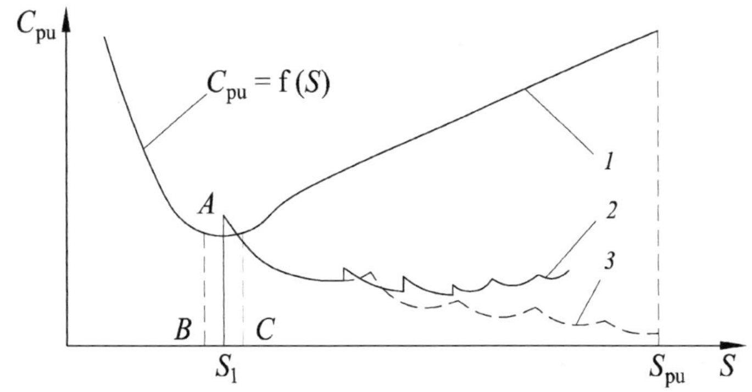

Figure 1. Reduced costs C_{pu} vs the length of a trench: S_1 – the length of the trench corresponding to the optimal service life of the cutters; S_{pu} – the length of the trench corresponding to the critical wear of the cutters limited by the strength of a brazed joint

Thus, the optimal life of the teeth depends on the numerical costs C_m and C_{re}, on the intensity of changes in the average feed rate of the trencher as the tool wear increases, and on some other restrictions.

In this case, the boundary conditions of tool durability [1, 5] are as follows: a) the strength of the brazed joint composed of a carbide plate and a tooth holder, which is limited by the critical tool wear in height; b) the absence of slipping of the undercarriage when the traction force for adhesion to the ground is greater than the traction force generated by the transmission and all the resistances that arise during the operation of the trencher.

These restrictions are interrelated, which means that as a tool wear area increases, the cutting resistance of frozen soil in the tangential and normal direction to the cutting path also increases. In adverse environments, the above situation entails a higher overall resistance to movement that can even make the machine stop. On the other hand, an increased wear surface of a cutting element leads to a significantly increased pressure from the frozen ground on the rear edge of the cutter. This pressure, along with a reduced brazed seam, greatly increases the likelihood to have the carbide insert detached from the tooth holder.

These restrictions are valid when trenchers cut homogeneous frozen soils that prevail in Siberia in winter [3].

When frozen soils containing solid inclusions are cut, it should be borne in mind that cutting tools might fail. The studies show [1] that the teeth break mainly when they cut gravel-pebble inclusions exceeding 100 mm in size.

4. The main results

In the light of the above, it is necessary to determine the optimal length of an excavated trench for various combinations of operating conditions.

The dependence of the actual feed rate v_a of the trencher on the length of the trench S_i (Figure 2) is almost linearly decreasing, and, with wear, the feed rate remains unchanged.

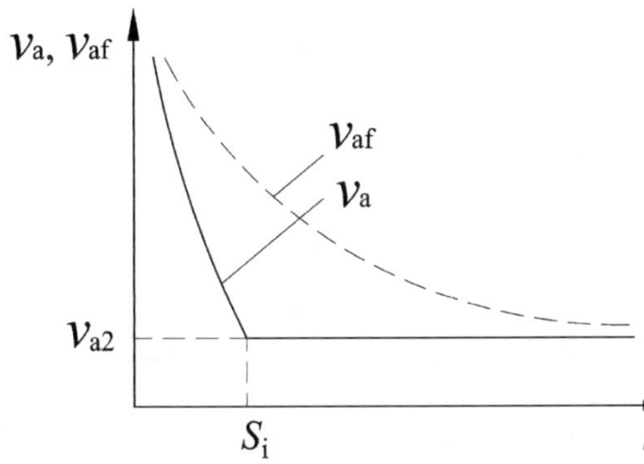

Figure 2. Changes in v_a and v_{af} indices vs trench length

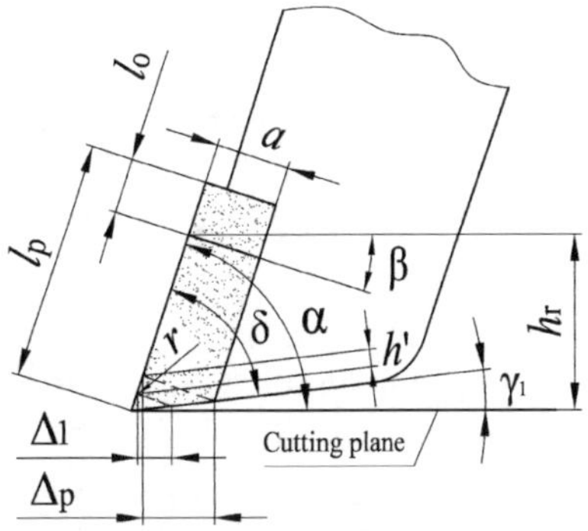

Figure 3. Scheme for determining the limit wear and the number of regrinds for a cutting element

The inflection point on the graph corresponds to the length of the excavated trench S_i, which is limited by the wear area of the cutter equal in plan to the thickness of the carbide cutter plate *a* (Fig. 3).

At S> Si, the feed rate is practically stabilized due to an insignificant increase in resistances on the wear surface of the cutter that are parallel to the cutting plane (Fig. 2).

With the power balance of the trencher engine [2], taking into account the wear of all working edges of the cutting element, we obtain the average feed rate depending on the length of the trench excavated during the period of wear formation less than or equal to the thickness of the carbide plate

$$v_{af} = \left[D_1 - D_4 \cdot S + \sqrt{(D_1 + D_4 \cdot S)^2 - 4S \cdot D_2 \cdot D_3} \right] / 2D_3, \quad (2)$$

where

$D_1 = N_e - A_1 \cdot A_2 \cdot v_r (B_{c2} + B_{k2})$; $D_2 = 0.5 v_r^2 \cdot K_{a1} \cdot I_p \cdot C_{gy} \cdot H_f / (l_c \cos\varphi_a)$; $D_3 = A_1 \cdot A_2 \cdot v_r (K_{c1} + K_{k1})$;

$D_4 = 0.5 v_r^2 \cdot K_{a2} \cdot I_p \cdot C_{gx} H_f / (l_c \cos\varphi_a)$; $A_1 = 0.85(1 + 0.55b)(1 - (90 - \alpha)/150) k_v k_e$;

$A_2 = 1 - w \cdot \mathrm{ctg}(\alpha + \varphi_a)$; $K_{c1} = [0.48 l_c \cos\varphi_a / (z_1 v_r)] \sum_{i=1}^{n_c} C_{ci}$; $B_{c2} = 0.373(t_p - b) \sum_{i=1}^{n_c} C_{ci}$;

$K_{k1} = [0.97 l_c \cos\varphi_a / (z_1 v_r)] \sum_{i=1}^{n_k} C_{ki}$; $B_{k2} = 0.24(t_t - b) \sum_{i=1}^{n_k} C_{ki}$; $K_{a1} = (\mu + w) \sum_{i=1}^{n_c + n_k} q_i$;

$K_{a2} = \mu \cdot C_x \dfrac{l_c \cos\varphi_a}{z_1 \cdot v_r} \sum_{i=1}^{n_c + n_k} q_i$.

Here N_e is the effective power spent on cutting frozen soils; v_r is the cutting speed; C_{gy} and C_{gx} are the coefficients reflecting the effect that geometrical parameters of teeth have on the width of wear surfaces of the rear and side faces, respectively; *b* is the width of the cutter; *α* is the cutting angle; k_v is the coefficient for cutting speed; k_e is the energy capacity factor; z_1 is the number of cutters in a digging chain; C_{ci} and C_{ki} is the strength of the ground according to the density meter in the zone of contact of the middle and extreme cutters with the ground; *μ* is the coefficient of resistance to wear surface movement; q_i is the pressure on the wear surfaces of the *i*-th cutter; C_x is the influence coefficient for the side surface wear on the cutting resistance; l_c is the length of the entire digging chain; φ_a is the angle of inclination of the working part of the cutter to the vertical; t_t is the step the cutters are placed on the chain; I_p is the intensity of tooth wear in height; H_f is the depth of the trench in a frozen layer; *w* is the coefficient of resistance to movement of the chain along the guides.

When the trencher cuts a trench of length *S* greater than the value of S_i, the average feed rate v_{af} will be

$$v_{af} = [v_{ap} S_i + v_{a2}(S - S_i)] / S, \quad (3)$$

where v_{ap} is the average feed rate of the trencher for the period when the wear area equal to Δ_p is formed; v_{a2} is the average feed rate of the trencher, which corresponds to the wear greater than the thickness of the carbide plate until the minimum permissible wear of the cutter in height l_0 is reached (Fig. 3).

The length of the excavated trench S_i is calculated as follows

$$S_i = (v_{ap} \cdot a \cdot l_c \cdot \cos\beta \cdot \cos\varphi_a)/(C_g \cdot H_f \cdot I_p \cdot v_r \cdot \sin(\alpha+\beta)), \quad (4)$$

where β is the angle between the wear surface and the cutting path; C_g is the influence coefficient for the cutter nomenclature on the wear width on the carbide edges.

The trench length S_1 corresponding to the minimum reduced costs is calculated by differentiating the equation (1) with respect to dS, equating the resulting derivative to zero, from the equation

$$dS_п/dS = \left((C_m + n_e \cdot K_c)/([v_{af}(S)] \cdot k_t) + C_{re}/S\right)' = 0. \quad (5)$$

It should be borne in mind that for low-strength and low-abrasive frozen soils, including clays and loams, the dependence of the reduced costs on the length of the excavated trench may not have a clearly expressed extremum. Therefore, the range of excavation of the trench S'_i is constrained by the durability of the brazed joint alone. Therefore, the length of the trench S'_i can be found by the formula

$$S'_i = S_i + [l_c \cdot \cos\varphi_a (h_r - \Delta_p/C_g) v_{a2}]/(H_f I_p v_r), \quad (6)$$

where h_r is the maximum wear height of the plate against the strength of the brazed joint (Fig. 3). It is found from the expression

$$h_r = [l_p - l_0 - a/\mathrm{tg}(\alpha+\beta)]\sin\alpha, \quad (7)$$

where l_p is the length of the carbide plate (Fig. 3).

When the reasonable trench length S_1 is less than the S_i value, the cutters should be resharpened. The number of permissible regrinds is calculated by the formula [5]

$$i_r = \frac{(l_p - l_0)\sin\delta - a \cdot \cos\delta}{r + h' + \Delta_1 \sin(\beta+\gamma_1)/\cos\beta} - 1, \quad (8)$$

where δ is the angle between the front and rear working surfaces of the cutter; γ_1 is the angle between the rear working surface of the cutter and the cutting plane; Δ_1 is the current wear of the cutter, not exceeding in terms of the thickness of the carbide plate; r is the radius of bluntness of the cutting edge; h' is the thickness of the removed layer of hard alloy for one regrinding.

Using the obtained dependencies, as applied to chain trench excavators, there are three main cutting options:

1. If the cutters are worn with excess Δ_p, the cutting tool should be replaced with a new one as soon as the trench length S_1 is reached (Point *A*, Figure 1);
2. If the wear does not exceed Δ_p, then regrinding of a tool is possible after excavating a trench of length S_1 (Graph 2, Figure 1). The number of regrinds is determined by the strength of the brazed joint;
3. A cutting element should be replaced with a reconditioned one as soon as the length of a trench corresponds to the minimum cost. In this case, the cutting tool should be replaced at the end of the work shift (Graph 3, Figure 1).

5. Conclusions

A cutting assembly should be replaced with a new one as soon as a trench of length S_1 is cut, particularly when a tooth life is reasonable (point *A*, Fig. 1). Operational calculations have shown that option 3 ensures the greatest economic efficiency (Graph 3, Fig. 1).

References

[1] Leshchiner V B 1991 *Improvement of Tools for Cutting Frozen Soils* (Tomsk: Publishing House of Tomsk State University)

[2] Leshchiner V B and Kravchenko S M 1986 *Methodology for Standardizing the Consumption of Trench Excavator Tools in the Development of Frozen Soils* (Tomsk Engineering and Construction Institute, Tomsk)

[3] Leshchiner V B and Kravchenko S M 1983 *Classification of Frozen Soils of Siberia and the North by Wear Capacity and its Application for Standardization of Tool Consumption and Regionalization of the Territory by this Indicator* (Tomsk Engineering and Construction Institute, Tomsk)
[4] Kravchenko S M and Slepchenko V A 2018 Determination of the Optimal Tool Life for Rotary Trench Excavators, *Mechanization in Construction* **5-6** 39-42
[5] Kravchenko S M and Slepchenko V A 2018 *Operation and Reliability of Construction and Road Machines. Textbook for universities* (Tomsk: Publishing House of Tomsk State Architectural and Construction University)

IOP Conf. Series: Materials Science and Engineering 1118 (2021) 012027 doi:10.1088/1757-899X/1118/1/012027

Efficient Technologies for Producing Cast Iron Billets and Products with Specified Properties and Microstructure

D A Lubyanoi[1], A V Markidonov[2], E Yu Pudov[1], V N Shakhmanov[1], E G Kuzin[1], and O S Semenova[1]
[1]Kuzbass State Technical University named after T. F. Gorbachev (a branch in Prokopyevsk), 19A Nogradskaya Str., Prokopyevsk, 653033, Russia
[2]Kemerovo State University (a branch in Novokuznetsk), Tsiolkovsky Str., Novokuznetsk, 23654000, Russia

E-mail: lubjanoy@yandex.ru

Abstract. The article discusses the use of resonant-pulsating refining technology in order to improve the quality of cast iron products. This technology is used in the cast iron production and cast iron mold induction. It is relatively simple and easily fits into existing production. The effectiveness of this technology is at the world's best indicator level. The chemical composition and smelting modes of cast iron influences significantly on the microstructure of gray cast iron. The determining factors are the content of silicon in cast iron, carbon, manganese, titanium and vanadium. The carbon content affects significantly taking into account its content above 3.8%, since this content increases the content of large, ripe forms of graphite in cast iron. Titanium and vanadium also influence significantly on the microstructure of cast iron. These two elements, even in small amounts, affect the size and shape of graphite greatly. It also affects the amount of ferrite, perlite and cementite in the structure of cast iron. Vanadium also affects greatly the service life of cast iron products.

1. Introduction

Metallurgy has a significant impact on key industries in the global economy. Many countries of the world are developing their own metallurgical production to meet domestic and external demand in the required range of metal products.

The main trend in the development of metallurgy and mechanical engineering is to reduce the material consumption and increase the durability of cast iron products. It is possible to increase the service life of products by changing the microstructure of cast iron using new high-performance technologies. The change in structure is achieved by micro-alloying cast iron transforming the size and shape of graphite inclusions. These inclusions increase the strength properties of the parts. It is also possible to use the technology of thermal processing and resonant-pulsating refining. These technological influences significantly increase the operational properties of gray cast iron products.

The introduction of modern resource- and energy-efficient technologies for the production of promising steels and cast iron remains relevant to meet the needs of economic sectors.

2. Formulation of the problem

It is necessary to determine the optimal content of alloying elements for the optimal technology for producing cast iron with the necessary properties. At the same time, it should be noted that optimal alloying could significantly reduce the content of manganese and silicon in heat-resistant cast irons.

The cast iron is industrially used for the manufacture of wear-resistant parts [1, 2] worldwide. The enterprises of Kemerovo region totally produce generally purposed grey cast iron of grades SCH-10 -

Content from this work may be used under the terms of the Creative Commons Attribution 3.0 licence. Any further distribution of this work must maintain attribution to the author(s) and the title of the work, journal citation and DOI.
Published under licence by IOP Publishing Ltd

SCH-30, grey cast iron of replacement steel casting equipment and mills for ferroalloys, grey half-cast iron of special purpose, and white alloyed cast iron of special purpose. The microstructure of grey and half-cast iron is shown in Figure 1 a), b).

The microstructure of gray cast iron is significantly influenced by the chemical composition and smelting modes of cast iron. The determining factors are the content of silicon in cast iron, carbon, manganese, titanium and vanadium. The carbon content affects greatly its content above 3.8%. Since this content increases the content of large, ripe forms of graphite in cast iron. It is mandatory to refine the metal in a ladle with nitrogen or argon in order to eliminate them or to conduct thermal treatment of the melt in an induction furnace. The influence of silicon also affects the microstructure of cast iron greatly. Thus, when in some cases its content is less than 0.8%, they could observe the appearance of cementite in the structure of cast iron. It is necessary to carry out either additional alloying of the metal in the furnace in order to eliminate it. Modifying the metal with ferrosilicon in the ladle is also necessary. Nitrogen purging also contributes to a favorable cast iron microstructure. Excessive silicon content in cast iron leads to an increase in ferrite in the structure of cast iron. This leads to a decrease in strength in cast iron and a decrease in the performance of cast iron products.

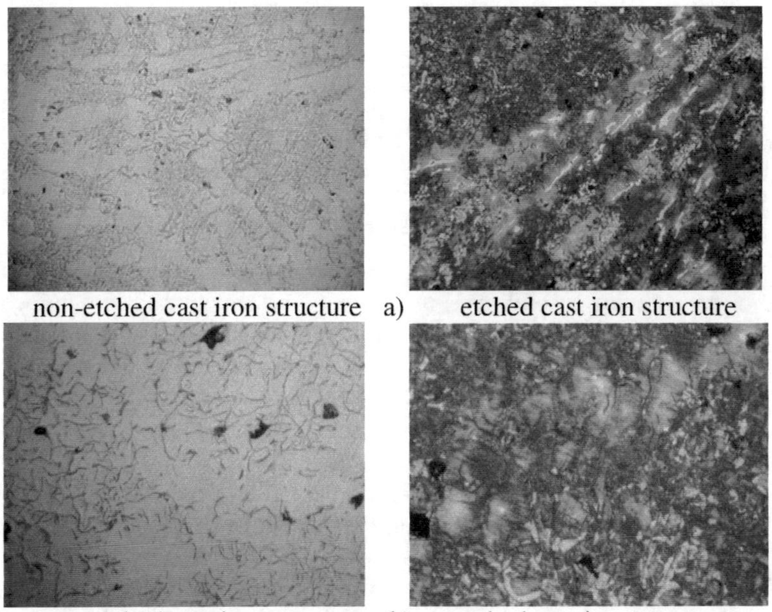

non-etched cast iron structure a) etched cast iron structure

non-etched cast iron structure b) etched cast iron structure

Figure 1. Influence of silicon content on the metal base and graphite phase of pallets (×100):

a) – Si-0.97%, V-0.04%, MEq.- 0.72%; b) – Si-1.02%,V-0.04%, MEq.- 0.59%.

It should be noted that the microstructure of cast iron is also greatly influenced by the mode of thermal treatment of the melt. For example, with a thermal exposure of 10 minutes, when overheating to a temperature of 1520-1550 ° C, mainly half-cast iron is formed (Figure 1, a). Heating the metal at the same exposure to temperatures of 1480-1520 ° C contributes to the formation of mainly gray cast iron, which has high strength characteristics and high thermal resistance, indicating as the world's best [3, 4]. The microstructure of cast iron is also significantly affected with the content of vanadium.

3. Research result

The microstructure of cast iron is also significantly influenced by titanium. These two elements, even in small amounts up to 0.1% significantly affect the size and shape of graphite and the amount of ferrite, perlite and cementite in the structure of cast iron. Vanadium also affects the service life of cast iron products greatly. In cast iron of Kemerovo region the vanadium content is usually in the range of 0.04-

0.07%. To select the optimal microstructure and chemical composition of cast iron with a vanadium content of 0.04-0.05% in cast iron and a structural diagram was developed. It allows choosing the optimal chemical composition of cast iron under the appropriate modes of thermal treatment of the melt (Figure 2).

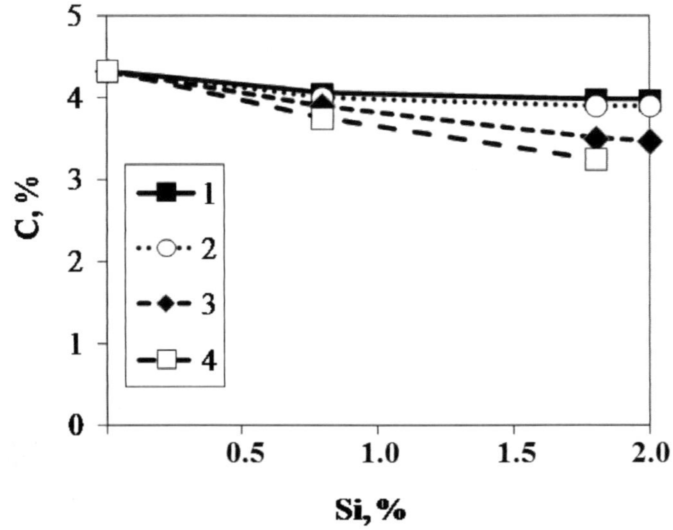

Figure 2. Structural diagram of grey naturally alloyed cast iron. The ranks of the limit region: 4-white cast iron, 3-half cast iron, 2-perlite gray cast iron, 1-perlite-ferritic cast iron (V – 0.04-0.05%).

White and half cast iron are used for producing rolls and other wear-resistant products. Perlite gray cast iron is used for manufacturing engineering products with increased structural strength (beams, brackets). Perlite-ferritic cast iron is used for the manufacture of heat-resistant products, in mechanical engineering and metallurgy (mills, pallets, intermediate bucket covers of continuous casting machines (CCM), machine beds).

Thus, setting the microstructure of cast iron based on the application of methods of thermal processing of the melt allows you to obtain special-purpose cast iron with the specified properties. The use of modern technologies for the cast iron production can significantly reduce the metal content of products. It can also increase their operational stability and reduce the cost of production.

In conclusion, it would be liked to note the following. As it follows from experimental studies, the micro hardness of phosphate eutectic in naturally alloyed titanium and vanadium blast furnace cast iron is 12-15%, which is higher than in cast irons that do not contain these elements [5-7]. Thus, alloying elements are able to neutralize the harmful effects of impurities. It is obvious that detailed research of this problem in the conditions of industrial production is not cost effective. Therefore, in this case, computer modeling is more rational. With the help of a computer modeling you can test theoretical developments, explain and predict phenomena that have not been fully covered yet by other research methods. In addition, the relative cheapness of data acquisition distinguishes the computer modeling. Therefore, a comprehensive study of this problem, which includes not only field experiments but also computer experiments, has been decided to conduct. There is a large volume of results of current computer modeling of binary Fe-C systems [8, 9]. For the Fe-P, Fe-S, Fe-V, and Fe-Ti systems, the volume of research is much smaller [10, 11], but there are studies of impurities of other elements.

For example, the interaction of impurities of various elements with grain boundaries in iron was studied in the work [12-15]. At the same time, the binding energy of clusters containing segregated atoms was calculated with the first-principle modeling. The interaction of impurities and alloying elements was also studied in the work [16]. The energy of dissolution of hydrogen atoms in an iron crystal is estimated as a function of the distance to the alloying elements. In General, it should be noted that in most works on modeling triple Fe-X-Y systems, one of the elements is usually hydrogen or carbon (see, for example, [17, 18]). For triple systems containing harmful impurities and alloying elements, such as Fe-V-P, the results of research are practically not found in both domestic and foreign literature.

The study resulted to the influence of the alloying elements Ti and V on the process of segregation of impurity atoms P and S. The optimal temperature conditions that promote migration of impurity atoms in the iron matrix were determined. The influence of P and S impurities on the strength properties of the iron polycrystal was evaluated.

Solving these tasks from a scientific point of view allowed us to deepen fundamental knowledge in the field of the influence of impurities on the properties of iron-carbon alloys. From the point of application, the research will improve technologies for updating the mechanical properties of structural materials. In addition, the approaches taken after the study allowed expanding the study and then switching for improving the wearing capacity of the replacing equipment like forging and bluming mills.

4. Conclusion

Thus, microalloying of cast iron with titanium and vanadium increases the strength properties of cast iron significantly thereby increasing the wear resistance of products operating in conditions of abrasive wear and high temperatures greatly. The optimal content of titanium and vanadium in gray cast iron is found to be in the range of 0.04-0.1%. This provides the required strength properties of the parts of the units increasing at the same time their heat resistance. Products made of this cast iron have the required wear resistance and can improve the equipment operational reliability. From a scientific point of view, the solution of these problems has significantly deepened the fundamental knowledge in the field of the impurities influence on the properties of iron-carbon alloys.

References

[1] Tsepelev V S, Selyanin I F, Lubyanoy D A and Baum B A 1995 *Steel* **5** 42
[2] Valeev D V, Zinoveev D V, Kondratiev A V, Lubyanoi D A and Pankratov D A 2020 *Metals* **10** 1
[3] Lubyanoi D A, Pudov E Yu, Kuzin E G, Semenova O S and Rybár R 2020 *E3S Web Conf.* **174** 03022
[4] Lubyanoy D A, Tolstikova Yu A, Markidonov A V, Kuzin E G and Buymov D V 2020 *Proceedings of higher educational institutions. Ferrous metallurgy* **63** 218
[5] Lubyanoy D A, Kustov B A, Novikov N I, Sinyavsky I A, Shulgin Yu f, Kokolevsky I V and Kalashnikova V V 2004 *Efficiency of modern ways to improve the quality of cast iron products and the development of metallurgical and machine-building enterprises in a competitive environment* (Novosibirsk: Publishing house of IEOPP SB RAS)
[6] Lubyanoi D A, Markidonov A V, Nevolin V M, Lubyanoi D D and Kambalin I O 2020 *IOP Conf. Series: Materials Science and Engineering* **848** 012048
[7] Lubyanoi D A, Shevchenko S Yu, Lubyanoi D D, Markidonov A V and Zvarych E B 2020 *IOP Conf. Series: Materials Science and Engineering* **848** 012049
[8] Becquart C S, Raulot J M, Bencteux G, Domain C, Perez M, Garruchet S and Nguyen H 2007 *Computational Materials Science* **40** 119
[9] Chirkov p V and Mirzoev A A 2013 *Bulletin of SUSU. Series "Mathematics. Mechanics. Physics»* **1** 114
[10] Wang B, Sak-Saracino E, Gunkelmann N and Urbassek H M 2014 *Computational Materials Science* **82** 399
[11] Ko W, Kim N J and Lee B-J 2012 *Journal of Physics: Condensed Matter* **24** 225002
[12] Mendelev M I, Han S, Son W, Ackland G J and Srolovitz D J 2007 *Physical Review B* **76** 214105
[13] Verkhovykh A V, Okishev K Yu, Mirzoev A A and Mirzaev D A 2016 *Bulletin of the South Ural state University. Metallurgy Series* **16** 32
[14] Hendy M, Hatem T M and El-Awady J A 2018 *TMS Annual Meeting & Exhibition*

[15] Kolesnikov V I, Migal Yu f, Doronkin V N, Novikov E S and Kolesnikov I V 2012 *Bulletin of the southern scientific center of the Russian Academy of Sciences* **8** 27
[16] Mirzoev A A, Mirzaev D A and Rakitin M S 2011 *Bulletin of SUSU. Series "Mathematics. Mechanics. Physics"* 77
[17] Burnyshev I N, Besogonov V V, Murin A V and Nagornykh I L 2007 *Bulletin of IzhSTU.* 120
[18] Kim H-K, Jung W-S and Lee B-J 2009 *Acta Materialia* **57** 3140

IOP Conf. Series: Materials Science and Engineering 1118 (2021) 012028 doi:10.1088/1757-899X/1118/1/012028

Failures of safety services in case of emergency fall of cargo-passenger elevator's cabin exploited in industrial and civil premises

M Popov and A Negodin
Tomsk State University of Architecture and Civil Engineering, Solyanaya Sq. 2, Russia

E-mail: semerka.82@mail.ru

Abstract.The growing fleet of elevators and other short distance vehicles (escalators, passenger conveyors and multi-cabin elevators) require constant improvement of installation methods and technical servicing to increase reliability and safety levels when used. When elevators are used for transporting people, the issue of reliability is becoming very important. The realibility indicator is not a constant value, since it has a direct dependence on a time factor, the main component of which is the standard service life of the main units and kinematic constrains of the equipment used for elevators. As a result, over time, the risk of accidents and emergencies in residential buildings may increase.

1. Introduction

An elevator has become an integral part of the artificially created human environment under technocratic civilization. For almost one century, it was possible to create a fully automated system used for internal transportation of passengers and goods in premises, which functions reliably without requiring special knowledge and preliminary training. The growing needs of social development require constant improvement of internal vehicles used in premises in line with cutting edge technologies. The growing fleet of elevators and other short distance vehicles (escalators, passenger conveyors and multi-cabin elevators) require constant improvement of installation methods and technical servicing to increase reliability and safety levels when used.

Some companies and organizations, which deal with elevator construction, installation and operation, consider as the key to susccessful solutions the qualified personnel who can keep up with the achievements of science and technology. The urgent need to have the elevator equipment used in premises against the backdrop of a dynamically developing industry was the reason for elevator constriction development.

At the end of the 19th century, hoisting winches with a rope pulley were created, in which the lifting force was created due to the frictional interaction of the ropes with the pulley rim. This invention served as a good basis for further improving elevators' design. It became possible to hang a cabin on several parallel branches of the traction ropes, which significantly increased the safety of passenger transportation. The risk of tightening a cabin under the shafts overlap was eliminated, since when the counterweight was mounted on a buffer, the tension of the ropes decreased, and their sliding stopped further cabin's lifting.

Content from this work may be used under the terms of the Creative Commons Attribution 3.0 licence. Any further distribution of this work must maintain attribution to the author(s) and the title of the work, journal citation and DOI.
Published under licence by IOP Publishing Ltd

IOP Conf. Series: Materials Science and Engineering 1118 (2021) 012028 doi:10.1088/1757-899X/1118/1/012028

The undoubted advantage of using the rope pulleys in the elevator's winches is the independence of their parameters from the vehicle lifting height. This opened up the broad prospects for mass serial production of elevators based on the principles of commoning and standardization of units and parts. The first samples of the rope pulleys had semicircular grooves in a rim, which did not provide the required traction for the ropes amd this fact limited the possibility of increasing the lifting capacity of an elevator's cabin.

To compensate for this shortcoming, in 1904, a winch design with a counter-sheave and a double girth of the rope pulley was developed. The double girth of the rope pulley has resulted in reduced rope durability. The solution was found through optimizing the shape of the cross-sectional profile of the grooves of the rope pulley rim providing by this the necessary traction capacity and durability of the ropes.

Further improvement of the elevator design was based on the fundamental decisions discussed above, subsequent achievements of science and industry. The history of the domestic elevator industry development reflected the world trends, although it was happening against the background of the planned economy features and isolation from the external market.

At the end of the 40s, the serial production of general-purpose elevators, special and high-speed elevators was manufactured in our country. In 1955-1956, VNIIPTMASH, together with the SOYUZLIFT Trust, created a standard range of passenger and freignt elevators. The systematic development of the domestic elevator industry is associated with the organization in 1963 of the Central Elevator Design Bureau (CEDB) and the further development of the corresponding production base in Moscow and other cities of the country.

The expansion of the production base was accompanied by specialized installation and operation service center for elevators, who was responsible for safety and reliability of the elevators. The companies such as SOYUZLIFTMONTAZH, MOSLIFT and others largely determined the performance quality of the domestic elevators.

Taking into account the foreign experience, in Russia there have been developed the State standards for various types of elevator equipment and architectural and planning solutions applied in constructing premises. Unfortunately, almost complete isolation from the world market and the lack of competition significantly hampered scientific and technological progress in the elevator industry.

Since the 90s, the situation has begun to change. The foreign competitors, when they appeared in the domestic market in the field of production and operation of elevators, have become the incentive to improve the technical level of domestic elevators in extremely unfavorable conditions, when the sharp production fall was in progress in the country.

Currently, there is a constant growth in the elevator fleet with a steady trend towards finding new design solutions that reflect the market requirements and scientific and technological advances in various industries. The organizational forms and technical means of the lift servicing are being improved. Serious attention is paid to increasing the productivity and quality of installation works.

Fierce competition in the domestic and world markets, expanding the range of customer needs in the elevator equipment serve as a good incentive to search for more effective technical solutions.

2. Main trends in elevator construction development

The following main trends in elevator construction development can be noted:

- using new building and finishing materials, including composite materials;
- improvement of cabins' construction and design as well as the equipment used for boarding places, taking into account the factor of anti-vandal resistance;
- design debugging of all systems involved in elevator equipment to reduce the level of noise and vibration in premises and in an elevator cabin;
- expansion of elevators' external installation in recessions of outwalls placed in residential and administrative premises of a tower-based type;
- reliability increase among devices that ensure the safe use of elevators;

- drive systems improvement and expansion of AC drives application with thyristor and amplitude-frequency control;
- control systems improvement based on the achievements of industrial electronics and microprocessor technology;
- use expansion of hydraulic plunger lifts with rope multipliers in low and medium-rise premises;
- comprehensive solution regarding the internal transport used in premises based on combined use of elevators, multi-chamber elevators, and escalators and passenger conveyors;
- extensive use of commoning and standardization methods to improve production quality, reduce the cost of mass production and operating costs;
- practice accumulation on modernization of the elevator equipment;
- service system improvement based on modern methods of computer information processing and control in combination with microprocessor self-diagnosis system for the elevator equipment;
- improvement of elevator design methods based on the widespread use of CAD;
- improvement towards the technology used in the elevator equipment manufacturing based on robotics for production processes;
- efficiency and quality rise towards installation of the elevator equipment based on advanced technology and robotics of labor processes.

To develop the infrastructure of the sectoral economy at industrial and civil construction facilities, the passenger elevators (hereinafter referred to as elevators) have long been used, which ensure people's transportation within the boundaries of the multi-storey space of houses and premises.

Complimented by the increase in multi-storey buildings construction development in cities and regions of our country, since the end of the 20th century, the elevator construction itself began to develop rapidly because the steady demand for the elevator equipment increased. This gave a significant impetus for manufacturers to initiate some improvement regarding the design and production of elevator vehicles. Also, the infrastructure responsible for certification bodies and elevators' control began to develop sharply.

The elevator evolution history says that an elevator was a vehicle that belonged to technical devices used at hazardous industrial facilities. The hazard class to these technical devices was also assigned, in addition, these were facilities under jurisdiction of the Federal Service of ROSTEKHNADZOR (Federal Sevirces for Environmental, Technological and Nuclear Supervision) of the Russian Federation. The hazard category depended mainly on people's movement in a passenger elevator cabin. At present, all responsibility for such vehicles' operation has reduced in certification bodies for technical devices that means the attention to processes and technologies used in production of these lifting vehicles has increased.

As a result, the list of manufactured elevator models has become wider. These models differ in design, security system, location of the main drive elements, automation in control, comfort and ergonomics, as well as, technical operating conditions. When elevators are used for transporting people, the issue of reliability is becoming very important. The reliability indicator is not a constant value, since it has a direct dependence on a time factor, the main component of which is the standard service life of the main units and kinematic constrains of the equipment used for elevators. As a result, over time, the risk of accidents and emergencies in residential buildings may increase. When such situations have emerged, they are impossible without human casualties, because people are the main users of passenger elevators.

The safety operation and reliability of the elevator in emergency situations, when cable binders have broken for the emergency braking, can be provided by the special constructions, called elevator catchers. Herewith, there is a certain risk of causing serious and bodily harm to people who are at that moment in a passenger cabin. As proof for it, there is a significant amount of information in mass media of the Russian Federation, as well as outside our country, on emergencies and their consequences.

According to the Committee on Analytics and Statistics (ELIAS) of the National Elevator Union, today in Russia there are about 530,000 elevators in apartment buildings, 150,000 of which have already worked their life cycle, which is 25 years. In other words, every third elevator needs to be replaced.

IOP Conf. Series: Materials Science and Engineering 1118 (2021) 012028 doi:10.1088/1757-899X/1118/1/012028

During this year about 25,000 elevators in new buildings have been put into operation in our country, and only 9518 elevators have been replaced, which is 2.2% of the required number. In 2020-2021, it is planned to replace about 5,000 old elevators, whereas 10,000 elevators are improving their resources every year. This means that the rate of equipment ageing, according to experts, will increase.

The technical condition of elevator equipment in Russia shows off, for last 5 years, the dynamics of fatal injuries due to not safety work of elevators:

in 2014 - 12 people were killed;
in 2015 - 12 people were killed;
in 2016 - 21 people were killed;
in 2017 - 35 people were killed;
in 2018 - 29 people were killed;
in 2019 - 32 people were killed.

These are officially recorded figures registered with the Federal Service of ROSTECHNADZOR. There are also unaccounted accidents that occurred when people moved in elevators. The reliability level decreases due to various factors: electrical, mechanical, physical, metrological and many others. This leads to necessity to carry out more detailed repair of the main elements of elevator equipment with documentary evidence and major repair of the hinged elevators. All these necessities are incentives to search for reliable safety constructions that will ensure for people a comfortable movement between floors.

References

[1] *Rules for the construction and safe operation of elevators*. PB 10-558-03
[2] Ioffe E Ya and Glumov V R 1994 *Elevator certification* (Moscow: Stroyizdat)
[3] Volkova D P 1999 *Elevators. Textbook for universities* (Moscow: ASV Publishing house)
[4] GOST R 53781-2010 *Elevators. Rules and methods of research (testing) and measurements for certification of elevators*. Sampling rules (M.: Standartinform, 2010)

IOP Conf. Series: Materials Science and Engineering 1118 (2021) 012029 doi:10.1088/1757-899X/1118/1/012029

Simulation on elevator emergency brake

M Popov and A Negodin
Tomsk State University of Architecture and Civil Engineering, Solyanaya Sq. 2, Russia

E-mail: semerka.82@mail.ru

Abstract. Modern calculation methods used for hoisting units are reduced to obtaining the value of the required braking force when the values of mass and speed have been captured. Copies of the oscillograms with kinematic parameters on braking of passenger elevator cabins show that the acceleration tendencies when landing on soft braking devices are not linear. Consequently, the cabin braking forces also do not change linearly and are not constant since the moment after braking has started to the moment when it stops. The calculation methods give only the maximum value of the required braking force and the length of the cabin braking distance. The elevator cabin braking (counterweight) with safety devices takes some time. During this time, the braking force changes from zero to the maximum value, which, at high speeds of the cabin (counterweight) affects the acceleration of slow-down process, and the brake-path length. When simulating the braking force of elevator catchers, it is necessary to take into account the change of the braking force value.

1. Introduction

Elevator catchers must stop and hold the downward-moving cabin (counterweight) on the guide ribs which are turned on by a speed limiter [1].

According to the operation principle, safety devices are a kind of linear shoe type brake, in which the pressure on the shoe is created due to the self-locking effect.

Smooth braking catchers by their design have a device that limits the force of normal pressure on brake shoes with a smooth surface. The braking force of such safety devices throughout all slow-down path usually remains constant that is determined by the design or the rules [3].

The limitation of the normal pressure force to a predetermined level is achieved by using a pre-compressed spring or an elastically deformable structural element.

The safety devices that are used for soft braking have different constructions, but none of them is perfect [2]. Pincers, due to their design flaws, do not provide a stable braking distance for the cabin. With several successive landings on safety devices, the gap among braking paths can reach 1-2 m, which can lead to overloads. Also, the pincers react unsatisfactorily to the instability of the guide head thickness, its waviness, deviation from flatness, and installation errors.

The braking force of soft braking devices is always time-dependent [2].

2. Calculation method for elevator braking forces

Modern calculation methods used for hoisting units are reduced to obtaining the value of the required braking force when the values of mass and speed have been captured. The equation for the cabin movement energy (counterweight), when landing on the safety devices is [3].

$$\frac{m_C \cdot V_H^2}{2} + m_C \cdot g \cdot S_T = K \cdot R \cdot S_T \qquad (1)$$

Content from this work may be used under the terms of the Creative Commons Attribution 3.0 licence. Any further distribution of this work must maintain attribution to the author(s) and the title of the work, journal citation and DOI.
Published under licence by IOP Publishing Ltd

where mC is the cabin captured mass (counterweight), kg; VH - the initial responding speed of safety devices; R is the braking force of the catchers, N; V - the speed of cabin movement (counterweight), m / s; ST - the braking path of the cabin (counterweight), m; g - gravity, m / s2; K is a coefficient that takes into account the dependence of the braking force on the braking path (K = 0.5 - according to the linear law of the braking force growth, K = 1 - with a constant braking force).
The formula to calculate the braking force is determined from the energy equation

$$R = \frac{m_C}{K} \cdot \left(g + \frac{V_H^2}{2 \cdot S_T} \right) \quad (2)$$

and the cabin braking paths (counterweight)

$$S_T = \frac{m_C \cdot V_H^2}{2(K \cdot R - m \cdot g)} \quad (3)$$

In this case, the acceleration of slow-down path will be determined by the braking force of the catchers.

$$a = \frac{R - m_C \cdot g}{m_C} \quad (4)$$

However, the braking process takes some time. During this time, the braking force changes from zero to the maximum value.

Copies of the oscillograms with kinematic parameters on braking of passenger elevator cabins [2] show that the acceleration tendencies when landing on safety devices of soft braking are not linear. Consequently, the cabin braking forces also do not change linearly and are not constant since the moment after braking has started until it stops. The calculation methods give only the maximum value of the required braking force and the length of the cabin braking path S_T.

When elevators are under certification, the safety devices must be checked. When testing safety mechanisms, in addition to other data, it is necessary to have a schedule on how the load is done on the elastic elements of the safety mechanisms and formulas or graphs showing the change in braking force depending on the parameter set for the safety mechanisms with smooth braking, which are designed to capture various masses [1,4]. The braking force developed by the safety devices when using the same type of guide ribs is determined as the average value of the average braking forces determined while tested. The braking force, developed by the safety devices with step or continuous regulation, is determined respectively for the minimum and maximum captured mass [1]. That is, they are also assumed to be equal to a constant value.

Modern high-speed elevators move at speed of 7-12 m / s. The mass captured by the catchers is 5000 - 10000 kg. The kinetic energy of a moving vehicle must be slowed down within 1-4 s. In this case, the cabin braking path should not exceed 15-20 m. When developing the safety devices that satisfy the given conditions, it is necessary to take into account the change of the braking force since the moment it starts until it ends. It is necessary to have a function of the normal pressure force of the catchers R = f (t) on the guide rib. Having such a dependence, it is possible to obtain the functions of the total energy of a moving vehicle E = f (R), speed V = f (E), acceleration A = f (V) and the cabin braking path S = f (V, A).

If the formula (1) is converted to the following form

$$E_{i+1} = \left(\frac{m_C \cdot V_i^2}{2} + m_C \cdot g \cdot S_i \right) - V_i \cdot R_i \cdot \varphi \cdot dt \quad (5)$$

where Ei is the total energy of a moving vehicle on i - a time interval, J; Vi - the cabin speed on i - a time interval, m / s; Si is the path done by the cabin for i - a time interval, m; Ri - the normal force of pressure on a brake shoe, N; φ is the coefficient of friction between the block and the guide rib, then for a sufficiently small value of dt and the given function R = f (t), we can obtain the dependences E = f (t), V = f (t), A = f (t), S = f (t).

$$V_{i+1} = \left(2 \cdot \frac{E_{i+1}}{m_C} \right)^{\frac{1}{2}} \quad (6)$$

$$A_{i+1} = \frac{V_{i+1} - V_i}{dt} \quad (7)$$

$$S_{i+1} = S_i + \left(V_{i+1} \cdot dt + \frac{A_{i+1} \cdot dt^2}{2} \right) \quad (8)$$

However, the braking path Si is sequentially calculated by equation (4) and will be calculated at the first stage of the iteration. Therefore, the braking path delta Si should be expressed in terms of the number of iterations that will be known at the zero step Ei and Vi, then

$$S_i = \left(V_i \cdot dt + \frac{A_i \cdot dt}{2} \right) \quad (9)$$

$$S_i = \left(\left(2 \cdot \frac{E_{i+1}}{m_C} \right)^{\frac{1}{2}} \cdot dt + \frac{\left(\left(2 \cdot \frac{E_{i+1}}{m_C} \right)^{\frac{1}{2}} - \left(2 \cdot \frac{E_{i-1}}{m_C} \right)^{\frac{1}{2}} \right) \cdot dt^2}{2 \cdot dt} \right) \quad (10)$$

The final equation for energy of the braking vehicle is written in the form

$$E_{i+1} = \left(\frac{m_C \cdot V_i^2}{2} + m_Л \cdot g \cdot \left(\left(2 \cdot \frac{E_{i+1}}{m_C} \right)^{\frac{1}{2}} \cdot dt + \frac{\left(\left(2 \cdot \frac{E_{i+1}}{m_C} \right)^{\frac{1}{2}} - \left(2 \cdot \frac{E_{i-1}}{m_C} \right)^{\frac{1}{2}} \right) \cdot dt}{2} \right) \right) - n \cdot V_i \cdot R_i \cdot \varphi \cdot dt \quad (11)$$

where n is the number of wear surfaces.

The friction of the catchers' working elements along the guide ribs (metal against metal) is unstable and depends on many factors, the main of which is the temperature in the contact zone of the wear surfaces. In addition, the value of the friction coefficient, taken into account when calculating the braking force of the safety devices, is determined approximately and varies throughout the entire process. When simulating the process of vehicle braking with the safety lifting devices, this phenomenon also needs to be taken into account.

If we represent it in the form of a polynomial the dependence of the friction coefficient on the sliding speed and the force of normal pressure as function of $\varphi = f(V, R)$, then equation (11) can be represented as

$$E_{i+1} = \left(\frac{m_C \cdot V_i^2}{2} + m_C \cdot g \cdot \left(\left(2 \cdot \frac{E_{i+1}}{m_C} \right)^{\frac{1}{2}} \cdot dt + \frac{\left(\left(2 \cdot \frac{E_{i+1}}{m_C} \right)^{\frac{1}{2}} - \left(2 \cdot \frac{E_{i-1}}{m_C} \right)^{\frac{1}{2}} \right) \cdot dt}{2} \right) \right) - n \cdot V_i \cdot R_i \cdot f(V,R) \cdot dt \quad (12)$$

The graphs of the braking forces and the cabin braking path depending on time are presented in Figure 2. They were obtained through simulation with formula (12). Initial data: the mass of the captured cargo M = 500 kg; the force of normal pressure of the block on the wedge in one case is constant from the second step of calculation (red line), in the other case it gradually increases (blue line); the response speed of the safety devices in both cases V = 1.2 m / s; the number of wear surfaces n = 4. As seen from the graphs the cabin braking path in the second case slightly exceeded S = 0.3m.

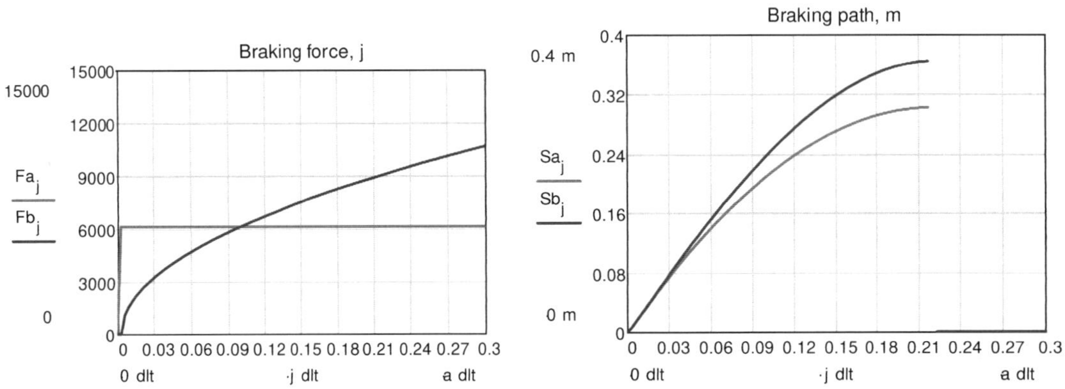

a) b)

Figure 1. a) graph of braking forces; b) graph of the dependence of cabin braking path on time

Under these conditions, the braking path length is obtained by formula (3) and it amounts to ST = 0.3m. Comparing these results, it can be noted that they are practically the same in terms of the braking path, but in the second case, a smooth increase in the braking force will lead to a decrease in dynamic loads on the elevator catchers and their drive, as well as the rest elevator system. By varying the braking force of the safety devices during emergency braking, it is possible to achieve the optimal technological characteristics in terms of dynamic effects.

3. Conclusions

1. The calculation methods used for the elevator safety devices are reduced to obtaining the maximum and minimum values of the braking force generated by the safety devices, and based on this minimum and maximum braking path of the elevator cabin (counterweight).
2. The elevator cabin braking process (counterweight) with the safety devices takes some time. During this time, the braking force changes from zero to the maximum value, which, at high speeds of the cabin (counterweight) affects the acceleration of slow-down process, and the brake-path length.
3. When simulating the braking force of the lifting catchers, it is necessary to take into account the change of the braking force value.

References

[1] *Rules for the construction and safe operation of elevators*. PB 10-558-03
[2] Ioffe E Ya and Glumov V R 1994 *Elevator certification* (Moscow: Stroyizdat)
[3] Volkova D P 1999 *Elevators. Textbook for universities* (Moscow: ASV Publishing house)
[4] GOST R 53781-2010 *Elevators. Rules and methods of research (testing) and measurements for certification of elevators*. Sampling rules (M.: Standartinform, 2010)

IOP Conf. Series: Materials Science and Engineering 1118 (2021) 012030 doi:10.1088/1757-899X/1118/1/012030

The Study of the Power Supply System of the Existing Gas Fields in Western Siberia using the Theory of Experimental Planning

I M Bogachkov

Tyumen branch of LLC "Gazprom engineering", 2, Vorovsky Street, 625000, Tyumen, Russian Federation

E-mail: bogim83@mail.ru

Abstract. Nowadays most of the gas fields in Western Siberia are at the final stage of development. To extend the cost-effective gas production at the final stage of development modern technologies are introduced at the fields. One of these technologies is the technology of distributed gas compression. As a result, a multiple increase in the electrical load is observed and new power transmission lines and substations are being constructed to provide electricity to gas fields. For example, at Vyngapurovsky gas field MCU (9 PCs) were to be put into operation and the 6 kV OHL with a length of 22 km were needed to be built, at Yamburg gas field at the moment MCU (52 PCs) are to be set up and it is required to build the 10 kV OHL with a length of 480 km. Thus, the purpose of this paper is to conduct an experimental study of the existing supply and distribution grids of the gas fields' power supply system in Western Siberia. To achieve this goal the following tasks have been solved: the method of the mathematical models development was selected; the main influencing factors on the voltage class were identified; variation ranges of the selected factors were defined; the optimization of the power supply schematic diagrams for the choice of voltage with minimal discounted costs was completed by the computer program PRON. The problems have been solved using the empirical method (measurement, comparison) of the study, the method of experiment planning (full factorial experiment), and computer programming methods. The analysis shows that all distribution grids of the power supply system in Western Siberia need to be converted to a higher voltage class of 20 kV or the additional overhead lines should be built.

1. Introduction

According to the Standart of Orfanization of Gazprom NTP 1.8-001-2004 [1], existing power supply systems for gas fields use a voltage class of 6/10 kV for a supply and distribution grid. At the end of the gas field's life, the estimated electrical load increases, which leads to the reconstruction of the external power supply system. Sometimes it is necessary to switch to a higher voltage class than 6/10 kV.

The peculiarity of the gas fields' power supply system makes it impossible to conduct real experiments on them. Therefore, changes in the parameters of these systems are carried out using mathematical modeling methods.

In the 80s, 90s and 2000s, the design of the power supply system for gas fields in Western Siberia was carried out without taking into account the electrical load at the final stage of development, which led to a brake on the introduction of modern technology for distributed gas compression [2-4].

The purpose of this paper is to perform an experimental study of the existing supply and distribution grids of the existing gas fields' power supply system in Western Siberia for the possibility of connecting MCUs to them.

Content from this work may be used under the terms of the Creative Commons Attribution 3.0 licence. Any further distribution of this work must maintain attribution to the author(s) and the title of the work, journal citation and DOI.
Published under licence by IOP Publishing Ltd

To achieve this goal, a number of tasks have been solved:

- the method of developing mathematical models was selected;
- the main influencing factors on the voltage class were determined;
- the ranges of variation of the selected factors were defined;
- the computer program PRON has been developed to determine the optimal voltage class at the minimum discounted cost;
- power supply schematic diagrams for voltage selection with minimal discounted costs in the PRON program were carried out.

2. Experimental section

The author has developed a software product PRON. It is a program for calculating the optimal stress class of gas fields, which takes into account its entire life cycle. PRON is based on the derived mathematical models of the supply and distribution grid of voltage classes 6, 10, 20, 35, 110, 220, 330 kV.

Table 1. The main levels and intervals of variation factors

Factor	Name of the factor	Basic zero level, x_{i0}	Range of variation, Δx	Upper level, «+»	Lower level, «-»
		Power supply grid – Main circuit with two through-lines			
		Gas field with higher electrical capacity (more than 75 MV)			
x1	Number of CGPP (PCs)	10.5	4.5	15	6
x2	Distance from the power source to the consumer L, km	105	45	150	60
x3	The coefficient of increase in the electric load k_{incr}, a.u.	3	2	5	1
x4	Load distribution coefficient for OHL k_{distr}, a. u.	0.775	0.225	1	0.55
		Gas field with average electric capacity (from 7 to 75 MV)			
x1	Number of CGPP (PCs)	3.5	1.5	5	2
x2	Distance from the power source to the consumer L, km	45	30	75	15
x3	The coefficient of increase in the electric load k_{incr}, a.u.	3	2	5	1
x4	Load distribution coefficient for OHL k_{distr}, a. u.	0.775	0.225	1	0.55
		Gas field with low electric capacity (up to 7 MV)			
x1	Number of CGPP (PCs)	4.5	2.5	7	2
x2	Distance from the power source to the consumer L, km	11.5	8.5	20	3
x3	The coefficient of increase in the electric load k_{incr}, a.u.	2.5	1.5	5	1
		Distribution grid – "Trunk circuit with one through trunk"			
x1	Number of gas clusters N, PCs.	6	4	10	2
x2	Distance from the power source to the consumer L, km	10.25	9.75	20	0.5
x3	The coefficient of increase in the electrical load during the period of falling production k_{incr}, a.u.	5.5	4.5	10	1
x4	Load distribution coefficient for OHL k_{distr}, a. u.	0.7	0.15	0.85	0.55

Mathematical models of the supply and distribution grids of gas fields were obtained on the basis of the theory of experimental planning. In the experiment, the object of research is the external power supply system of a gas field. The method consists in planning of the full factorial experiment (FFE) of type 2k, where k is the number of considered factors. All possible combinations of factor levels are implemented in the experiment. The parameter for optimizing the external power supply system is the voltage class.

The experiment is carried out by changing the factors of the power supply system, resulting in the optimal voltage class. The optimal voltage is determined by the minimum discounted cost. The optimal voltage class with the minimum discounted costs is found by changing the factors of the power supply system and determining the discounted costs for each voltage class.

For each influencing factor, in Table 1, the ranges of change (variation) are selected, allowing to cover a significant number of power supply schematic diagrams of the complex gas preparation plants (CGPP) and gas well clusters.

The determination of the coefficient of increase in the electric load for the supply and distribution grid was given by the author in the article "The Assessment of the dynamics of the electric load in the gas fields of Western Siberia, taking into account the entire life cycle".

The coefficient of increase in the electrical load is:

1. For the power supply grid:
- in the first period of a gas field's life - k_{incr} = 1a.u.;
- in the second period of a gas field's life - k_{incr} = 1.63 a.u.;
- in the third period of a gas field's life - k_{incr} = 3.45 a.u.

2. For the distribution grid:
- in the first/second period of a gas field's life - k_{incr} = 1 a.u.;
- in the third period of a gas field's life – k_{incr} = 10 a.u.

3. Results

With the help of PRON program, the existing supply and distribution grids of gas fields in Western Siberia have been tested according to the criterion of optimal voltage class, taking into account the coefficient of increase in electrical load in different periods of the field's life cycle. Table 2 shows the parameters of the existing supply grid of a number of gas fields in Western Siberia.

Table 2. Technical characteristics of the existing grid at the CGPP of the existing fields of Western Siberia.

#	Name of the power transmission line	Number of CGPP, PCs	The length of the trunk line, km	Accepted voltage class, kV	Optimal class voltage, kV		
					the first period kincr = 1	the second period kincr = 1.63	The third period kincr = 3.45
	The Fifth section of the Achimov deposits, CGPP 51						
1	OHL SS 6/35 kV – SS 35/6 kV CGPP 51	1	8.95	35	20	35	35
	Vyngapurovsky gas field						
1	OHL Vyganpur – Peschanaya	1	25.71	110	20	35	110
	Komsomolsky gas field						
1	OHL Tarko - Sale CGPP	-	15.26	110	20	35	35
	Medvezhye gas field						
1	OHL Bazovaya-PGF 9-1,2	10	111.81	110	110	110	110
	Yamburg gas field						
1	OHL Yamburg	10	104.8	110	110	110	110

where kincr is the coefficient of increase, the definition of which is given in the article "The Assessment of the dynamics of the electric load in the gas fields of Western Siberia, taking into account the entire life cycle"

Table 3 shows the estimation of discounted costs for each supply grid of the external power supply system from Table 4 for voltage class 6, 10, 20, 35, 110 kV (for low-power gas fields).

Table 3. Calculated values of discounted costs of existing supply lines of the existing fields' power supply system in Western Siberia (low power).

Period of life	Name of the power transmission line	Discounted costs for the voltage class, billion rub.				
		6 kV	10 kV	20 kV	35 kV	110 kV
Gas fields with low electric power (up to 7 MV)						
The Fifth section of the Achimov deposits, CGPP 51						
1	OHL 35 kV SS 6/35 kV – SS 35/6 kV CGPP 51 I l, p	1.9	0.9	0.5	0.6	1.2
2	OHL 35 kV SS 6/35 kV – SS 35/6 kV CGPP 51 II l, p	3.3	1.5	0.8	0.8	1.3
3	OHL 35 kV SS 6/35 kV – SS 35/6 kV CGPP 51 III l, p	7.5	3.1	1.6	1.2	1.5
Vyngapurovsky gas field						
1	OHL 1 Vyganpur – Peschanaya I l,p	4.8	2.1	0.9	1.0	1.7
2	OHL 1 Vyganpur – Peschanaya II l,p	8.7	3.5	1.6	1.2	1.8
3	OHL 1 Vyganpur – Peschanaya III l,p	20.1	7.5	3.3	2.1	2.1
Komsomolsky gas field						
1	OHL 1 Tarko - Sale CGPP I l,p	2.9	1.3	0.75	0.8	1.4
2	OHL 1 Tarko - Sale CGPP II l,p	5.2	2.2	1.1	1.0	1.5
3	OHL 1 Tarko - Sale CGPP III l,p	11.9	4.6	2.1	1.4	1.7
(shaded)	- minimum discounted costs					

Table 4. Calculated values of discounted costs of existing supply lines of power supply system at operating gas fields in Western Siberia (medium and high power).

Period of life	Name of the power transmission line	Discounted costs for the voltage class, billion rub.				
		20 kV	35 kV	110 kV	220 kV	330 kV
Gas fields of average electric capacity (from 7 to 75 MV)						
Medvezye gas field						
1	OHL 1 Bazovaya-PGF-9-1,2	41.2	19.7	11.7	19.6	33.1
2	OHL 1 Bazovaya-PGF-9-1,2	62.6	29.0	13.9	20.4	34.1
3	OHL 1 Bazovaya-PGF-9-1,2	124.4	56.0	20.3	22.8	37.1
2-1	OHL 1 Bazovaya-PGF-9-1,2 The difference in the costs of the second life period from the first one	21.4	9.3	2.2	0.8	1
3-1	OHL 1 Bazovaya-PGF-9-1,2 The difference in the costs of the third period of life from the first one	83.2	36.3	8.6	3.2	4
Gas fields with large electrical capacity (more than 75 MV)						
Yamburg gas field						
1	OHL Yamburg the first period of life	37.5	17.9	11.4	19.0	32.0
2	OHL Yamburg the second period of life	56.6	26.3	13.5	19.8	33.0
3	OHL Yamburg the third period of life	111.8	50.6	19.7	22.1	36.0
2-1	OHL Yamburg The difference in the costs of the second life period from the first one	19.1	8.4	2.1	0.8	1
3-1	OHL Yamburg The difference in the costs of the third period of life from the first one	74.3	32.7	8.3	3.1	4
(shaded)	- minimum discounted costs					

There is the estimation for voltage classes 20, 35, 110, 220, 330 kV (for medium-capacity gas fields) in Table 4.

Thus, analyzing Table 4, it can be concluded that the higher the voltage class, the smaller the difference in discounted costs. It is seen that for the 6 kV voltage class, the difference in discounted costs in the subsequent life cycle periods is large relative to the first period, and for 110 kV, the difference is smaller and then it converge to almost zero.This suggests that the higher the voltage class, the more the supply grid of the external power supply system is adapted to the growth of the electrical load, i.e., to the development of the gas field.

Table 5 shows the parameters of the existing distribution grid (the number of gas well clusters, the length of the main line, the accepted voltage class for electricity transmission) of some existing gas fields in Western Siberia. With the help of PRON program, the optimal voltage class of the existing distribution grid was selected for each life period (taking into account the load gain coefficient), the results are presented in Table 5. Table 5 shows that the most progressive voltage class for the distribution grid of gas fields is 20 kV.

Table 5. Technical characteristics of existing power distribution lines for gas well clusters of operating fields in Western Siberia.

#	Name of the power transmission line	Number of clusters, PCs	Power trunk length, km	Accepted voltage class, kV	Optimal voltage class, kV		
					The first period kincr = 1	The second period kincr =1	The third period kincr = 10
	The Fifth section of the Achimov deposits, CGPP 51						
1	OHL Cluster K-5A22	6	9.79	10	20	20	20
2	OHL Cluster K-5A23	7	16.4	10	20	20	20
3	OHL Cluster K-5A24	5	12.94	10	20	20	20
4	OHL Cluster K-5A25	5	10.49	10	20	20	20
	The First section of the Achimov deposits, CGPP 31						
1	OHL Cluster 1	8	15.62	6	20	20	20
2	OHL Cluster 3	6	10.28	6	20	20	20
3	OHL Cluster 18A	4	10.96	6	20	20	20
4	OHL Cluster 32	11	15.19	6	20	20	20

Thus, analyzing the values of discounted costs, it can be concluded that the voltage class of 20 kV is optimal for the distribution grid of the gas well clusters' external power supply system. With a tenfold increase in the electrical load, the difference between the values of the discounted costs is minimal. This means that the distribution grid with a voltage of 6/10 kV is not adapted to increase of the load by 10 times (there is no reserve) and additional capital costs are required for power supply. The values of the discounted costs of the grid with a voltage of 6/10 kV with an increase in the electrical load become comparable to the values of the discounted costs for the voltage class 35 and 110 kV. For the distribution grid of gas well clusters, the minimum discounted costs are inherent in the 20 kV voltage class.

Table 6 shows the results of calculations of discounted costs for each distribution grid of the power supply system from Table 5 for the voltage class 6, 10, 20, 35, 110 kV.

IOP Conf. Series: Materials Science and Engineering 1118 (2021) 012030 doi:10.1088/1757-899X/1118/1/012030

Table 6. Estimated values of discounted costs of existing power distribution lines for gas well clusters in existing fields of Western Siberia.

#	Name of the power transmission line	Discounted costs for the voltage class, million rub.				
		6 kV	10 kV	20 kV	35 kV	110 kV
	The Fifth section of the Achimov deposits, CGPP 51					
	The first and the second periods of life					
1	OHL Cluster K-5A 22	143.03	138.83	135.1	215.85	408.61
2	OHL Cluster K-5A 23	227.43	217.56	211.52	321.3	565.42
3	OHL Cluster K-5A 24	174.42	170.06	165.62	249.17	470.04
4	OHL Cluster K-5A 25	145.99	142.71	139.04	214.35	415.27
	The third period of life					
1	OHL Cluster K-5A 22	603.99	374.13	279.64	340.69	841.79
2	OHL Cluster K-5A 23	1116.33	688.07	444.08	491.72	1063.87
3	OHL Cluster K-5A 24	679.54	414.01	296.26	353.67	826.67
4	OHL Cluster K-5A 25	557.19	341.57	256.94	315.23	771.6
	The First section of the Achimov deposits, CGPP 31					
	The first and the second periods of life					
1	OHL Cluster 1	224.36	213.11	207.06	321.68	556.98
2	OHL Cluster 3	148.81	144.34	140.46	222.82	419.57
3	OHL Cluster 18A	146.06	143.93	140.39	209.58	416.77
4	OHL Cluster 32	231.53	216.9	210.47	338.5	565.38
	The third period of life					
1	OHL Cluster 1	1191.59	739.34	473.38	519.44	1128.24
2	OHL Cluster 3	632.19	391.04	288.41	348.79	852.35
3	OHL Cluster 18A	490.02	296.44	228.4	285.2	697.33
4	OHL Cluster 32	1405.93	880.08	551.9	593.4	1281.48
	(shaded) - minimum discounted costs					

4. Discussion

The experimental study of existing power supply systems has showed:

1. The number of gas well clusters in the distribution grid varies from 2 to 15 PCs. The length of the power transmission line of the distribution grid is from 3 to 16 km.
2. The optimal voltage class of a distribution grid is 20 kV.
3. The optimal voltage class of the supply grid of low-power gas fields (up to 7 MV), taking into account the increase in loads, is 35 kV and 110 kV;
4. The optimal voltage class of the supply grid of medium-capacity gas fields (from 7 to 75 MV) is 110 kV.

5. Conclusion

The voltage class at all the considered operating supply and distribution grids of gas fields in Western Siberia is not optimal, and the implementation of distributed gas compression technology will require additional capital investments for the construction of new power transmission lines and substations.

Thus, for the external power supply system of new gas fields in Western Siberia, it is recommended to use a voltage class of 110 kV for electricity transmission, and for a distribution grid it should be 20 kV. These voltage classes have a sufficient reserve of power for the transmission of electricity, taking into account the entire life cycle of the field (the growth of the electric load) at the lowest discounted costs.

References

[1] The Standarts of Organization of Gazprom NTP 1.8-001-2004 Norms of technological design of objects of gas-producing enterprises and underground gas storage stations of LLC "VNIPIgazdobycha" 2004 Moscow

[2] Sarancha A V and Ogay V A 2016 Application of modular compressor units of Russian production in the production of low-pressure Cenomanian gas *Academic Journal of Western Siberia* Vol 12 **1 (62)** pp 21-22

[3] Minlikaev V Z, Dikamov D V, Arno O V *et al* 2015 Application of mobile compressor units at the final stage of development of gas deposits *J. Gas industry* **1 (717)** pp 15-17

[4] Arno O B 2015 Technical and technological solutions and innovations at different stages of the life cycle of Yamburg deposits *Sc. J. of the Russian newspaper society* **2-3** pp 7-14

Developing the calculation methods of effective values of current and voltage for nonsinusoidal transient modes in electric power systems based wavelet transform

O V Arkhipova, N N Dolgikh, V Z Kovalev, D S Osipov, A O Paramzin and V A Tkachenko

Oil and Gas institute in Yugra State University, 16 Chekhov str., Khanty-Mansiysk, 628011, Russia

E-mail: d_osipov@ugrasu.ru

Abstract. The necessity to develop analysis and calculation methods of nonsinusoidal modes in electric power networks is conditioned by the increasing number of electrical receivers parental to non-linear distortions (variable-frequency electric drive, LED-based lightning and so on). Higher harmonic components increase power and efficiency loss in current carrying parts, result in overheating of conductors and intense insulation deterioration, can cause resonance overvoltage and automation equipment malfunctions. Due to digitalization of the electric power industry it is necessary to develop methods of receiving, processing and digital communication on system state variables (currents, voltage, power), including nonsinusoidal modes. Wavelet transform is wider applied to analyze complex harmonic processes in electric power industry in conditions of transient modes. The paper introduces calculation methods of integrated indexes of electric power system modes – currents, voltage and power in the presence of electrical receivers with non-linear current-voltage characteristics. The method is based on Parseval equality and allows calculating via wavelet coefficients received by transformation of the initial stream of transient values of currents and voltages. Herewith the issue of data compressing is solved when transmitting digital information. The method allows determining the variables of every frequency component, interpreting the data in the three-dimensional scale (peak values, frequency and time) which will be used in choosing activities for higher harmonics filtration to reduce the loss in current carring parts.

1. Introduction

Fourier processing is conventionally used to harmonically analyze nonsinusoidal processes in electrical power industry. Fourier processing is applied for the spectral analysis of resonance oscillations with the limiting amplitude in a mechanical system of vibration electromagnetic activators [1]. This method had a restriction; it was applicable only to sinus wave oscillations of the system with relatively low damping, which is why works [2, 3] suggest improvements to the method. When analyzing transient modes a negative effect of the spectrum flow appears, which can be reduced by Short Time Fourier Transformation. Nevertheless, due to Heisenberg equivocating, it is impossible to achieve good frequency and time resolution. To reduce the effect of the spectrum flow, the work [4] introduces a “pseudowavelets” method. To analyze complex harmonic transient processes in electric

Content from this work may be used under the terms of the Creative Commons Attribution 3.0 licence. Any further distribution of this work must maintain attribution to the author(s) and the title of the work, journal citation and DOI.
Published under licence by IOP Publishing Ltd

power supply industry a mathematical apparatus of wavelet transform is more often used. This kind of transform is widely used to analyze the parameters of electric power quality, including harmonic and interharmonic analysis [5, 6], transient simulation, short circuits and ground faults [7], electric equipment diagnostics and others. Decomposition of the data rows, received as a result of physical tests (measurements), applying the wavelet transform theory allows determining frequency ranges of the required width in the signal (depending on the sampling rate of the initial signal and decomposition depth), allows determining time and occurrence of some harmonic component.

2. Basic theorems of wavelet transform theory

There are continuous and discrete wavelet transforms in scientific and technical literature. When solving practical tasks connected with data digital processing, continuous wavelet transform is excessive. In its turn, discrete wavelet transform due to its peculiarity, when further wavelet decomposition is done only by approximate wavelet coefficients, is also restricted in its application. A variety of discrete wavelet transformation is a batch transformation, done according to the scheme in figure 1.

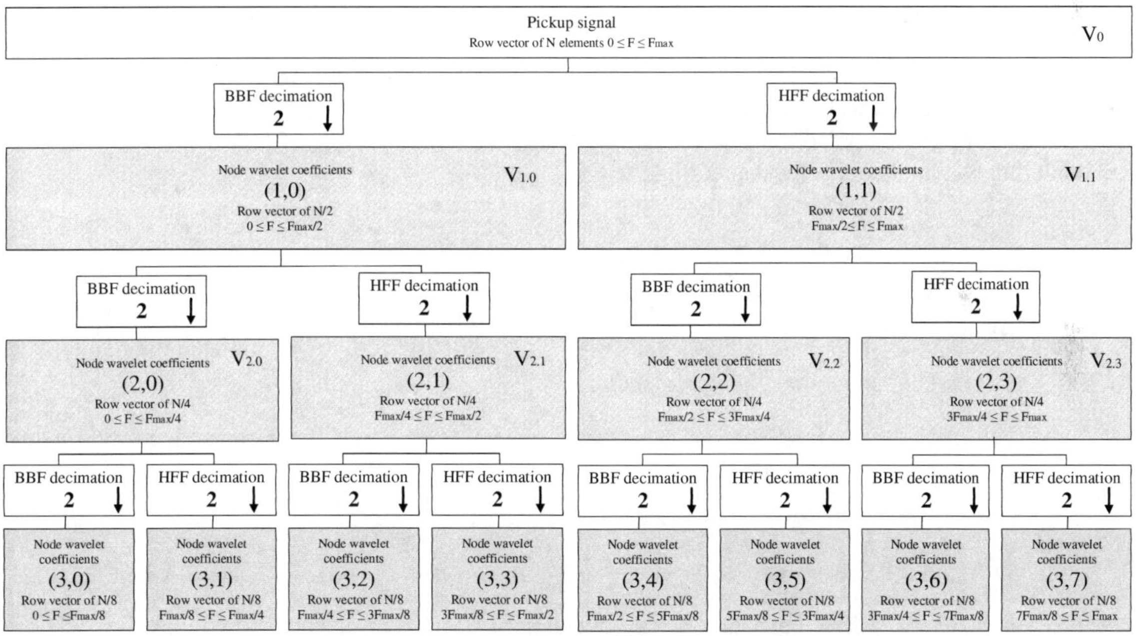

Figure 1. A scheme of batch wavelet transform.

Batch wavelet transform allows more evenly localization of frequency ranges of a signal, which is convenient for analyzing nonsinusoidal transient modes of electric power systems. Nowadays, there are fairly enough wavelet functions, which have different gain-frequency characteristics. For effective application of wavelet transform for calculating integral characteristics of a nonsinusoidal mode it is necessary to formulate a choice criterion for an optimal wavelet function. Two conditions can be used as a criterion: 1) minimum error when desampling by the wavelet coefficients; 2) uppermost energy of the spectrum for upper harmonic, corresponding to the pass band of wavelet function (wavelet filter). In this case, the choice criterion of the wavelet function can be written as an equation (1).

$$\begin{cases} \sigma = \sqrt{\dfrac{\sum\limits_{k=1}^{n}\left(i_k - \hat{i}_k\right)^2}{\sum\limits_{k=1}^{n} i_k^2}} \rightarrow \min; \\ \Delta_{En} = \dfrac{E_n}{E_\Sigma} \dfrac{\sum\limits_{k=n}\left|i_{j,k}^A\right|^2 + \sum\limits_{k=n}\left|i_{j,k}^D\right|^2}{\sum\limits_{k=-\infty}^{\infty}\left|i_{j,k}^A\right|^2 + \sum\limits_{j=J}^{\infty}\sum\limits_{k=-\infty}^{\infty}\left|i_{j,k}^D\right|^2} \rightarrow \max. \end{cases} \tag{1}$$

A choice of the optimal wavelet function according to (1) allows increasing determination accuracy of the effective value of current and voltage of individual harmonic.

3. Methods for determining the effective value of current and voltage of individual harmonic by wavelet coefficients

Determining higher harmonic for loss analysis and elaborating measures on filtration is a relevant task [8]. Increase of LED-based lighting results in overload of current-carrying parts with higher harmonic [9] and onset of voltage oscillations and a flicker [10]. Wavelet transform allows calculating integrated indexes of the electric power system mode for every frequency component. According to the theory of wavelet transform, instantaneous value of voltage and current, recorded by digital instruments can be represented by wavelet coefficients according to the formulas:

$$i(t) = \sum_{2^{-j}nm}^{2^{-j}n(m+1)-1} i_{j,m}(k)\cdot\psi(t) \tag{2}$$

$$u(t) = \sum_{2^{-j}nm}^{2^{-j}n(m+1)-1} u_{j,m}(k)\cdot\psi(k) \tag{3}$$

Active power can be determined for every frequency range via wavelet coefficients according to the scheme of decomposition (figure 1):

$$P_{j,m} = \frac{1}{n}\left(\sum_{2^{-j}nm}^{2^{-j}n(m+1)-1} i_{j,m}(k)u_{j,m}(k)\right) \tag{4}$$

Electrical conductance will be determined via wavelet coefficients of currents and voltages

$$g = \frac{P_{j,m}}{(U_{j,m})^2} = \frac{\frac{1}{n}\left(\sum\limits_{2^{-j}nm}^{2^{-j}n(m+1)-1} i_{j,m}(k)u_{j,m}(k)\right)}{\frac{1}{n}\sum\limits_{2^{-j}nm}^{2^{-j}n(m+1)-1} u_{j,m}^2(k)} \tag{5}$$

According to the theory of Frise reactive power, active current recorded via wavelet coefficients

$$i_a = gu_{j,m} = \frac{\frac{1}{n}\left(\sum\limits_{2^{-j}nm}^{2^{-j}n(m+1)-1} i_{j,m}(k)u_{j,m}(k)\right)}{\frac{1}{n}\sum\limits_{2^{-j}nm}^{2^{-j}n(m+1)-1} u_{j,m}^2(k)} \cdot \frac{1}{n}\sum_{2^{-j}nm}^{2^{-j}n(m+1)-1} u_{j,m}(k) \tag{6}$$

Effective value of the active current in this case will be determined by the formula:

$$I_a = \sqrt{\frac{1}{2^N} \sum_{2^{-j}nm}^{2^{-j}n(m+1)-1} i_a^2(k)} = \sqrt{\frac{1}{2^N} \sum_{2^{-j}nm}^{2^{-j}n(m+1)-1} i_{j,m}^2(k)} \quad (7)$$

This method allows determining effective value of current and voltages of individual harmonic. At the same time wavelet coefficients allow determining time (occurrence length) of every harmonic component.

4. Simulation modeling. Numerical experiment

A simulation model of the electric power system 110/35/6 kv was developed to test this method in Matlab Simulink software (figure 2). A control station for immersible pumps was determined as a node for non-linear load. A spectral composition of non-linear load was determined with certified equipment (Metrel MI 2792A).

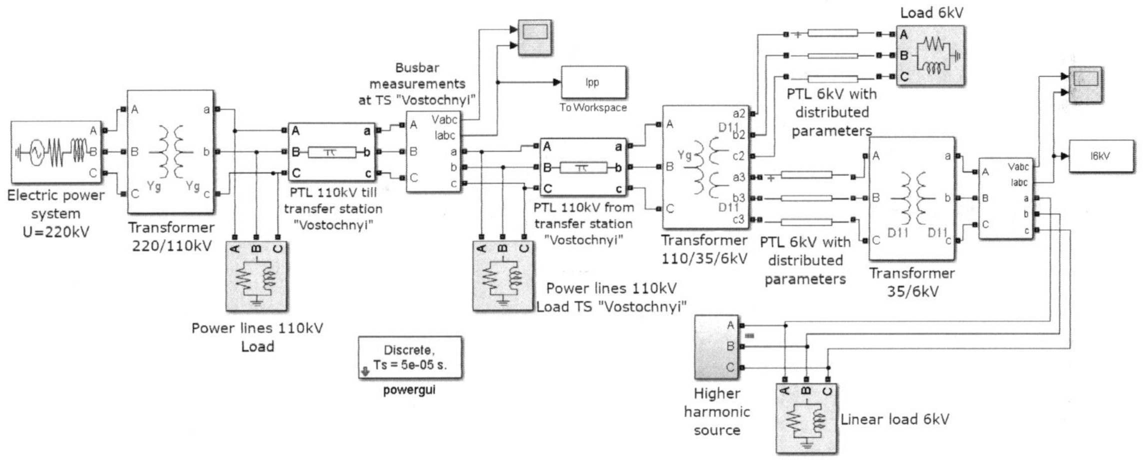

Figure 2. Simulation model to analyze a mode of the electric power system.

As a result of wavelet transform of the bus voltage in substation 6 kv (figure 3) a set of wavelet coefficients was received (figure 4).

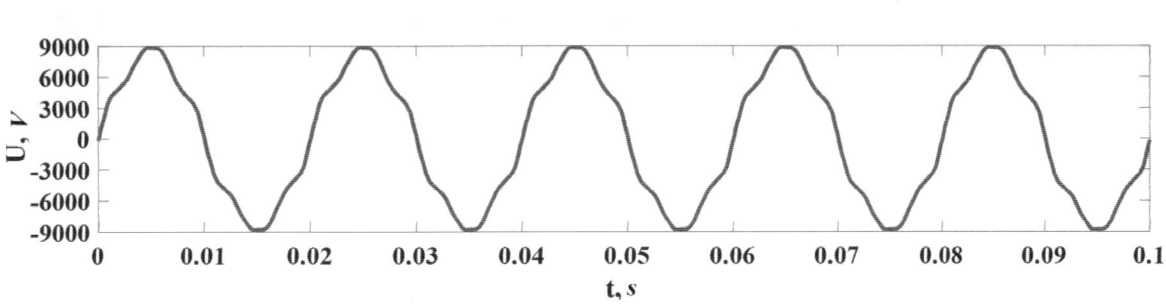

Figure 3. Voltage signal, subject to batch wavelet transform.

The analogous procedure of wavelet transform, according to the scheme (figure 1) was conducted for a range of instantaneous current.

IOP Conf. Series: Materials Science and Engineering 1118 (2021) 012031 doi:10.1088/1757-899X/1118/1/012031

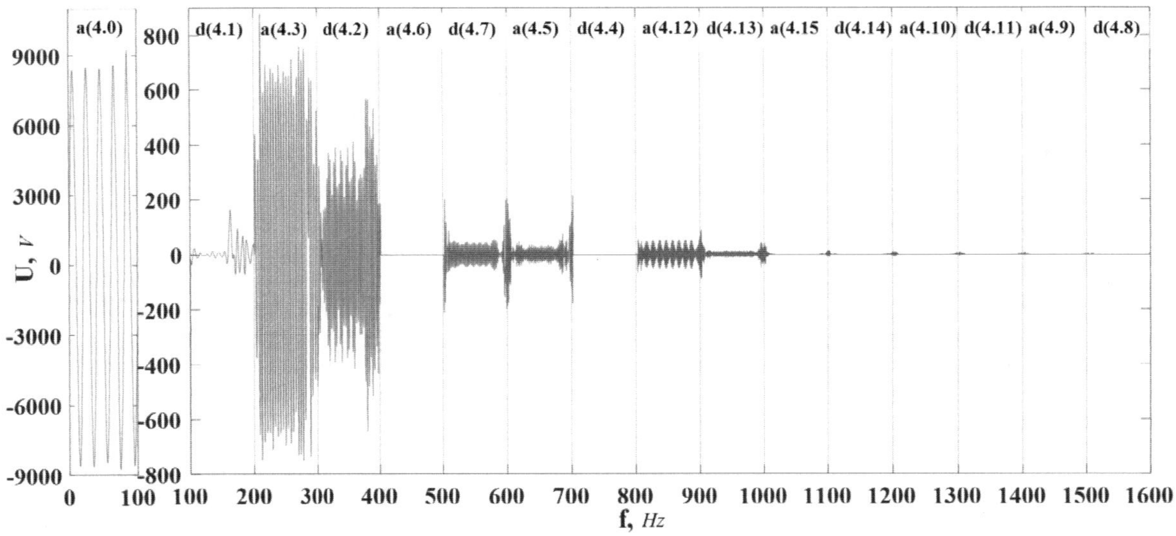

Figure 4. Result of batch wavelet transform of the voltage signal.

With this method active, reactive and distortion power have been determined for every harmonic recorded with the measuring instrument (table 1).

Table 1. Calculation of mode parameters by wavelet coefficients.

	Frequency, Hz	Symbol, measurement unit	True value	Value, calculated via wavelet coefficients
Active current component	50	I_{a1}, A	523	523
	250	I_{a5}, A	89	90
	350	I_{a7}, A	55	56
Actual power	50	P_1, kW	5 701	5 694
	250	P_5, kW	970	978
	350	P_7, kW	597	602
Reactive power and distortion power	50	$Q_1 + N_1$, kVar	4 103	4 105
	250	$Q_5 + N_5$, kVar	728	727
	350	$Q_7 + N_7$, kVar	471	468

This method can be applied to calculate interharmonic components, if the scheme of decomposition will correspond to accentuation of this frequency band with application of additional algorithms of interharmonic localization [11].

5. Conclusion

Applying mathematical apparatus of wavelet transform to calculate transitional nonsinusoidal modes of electric power systems allows eliminating a negative effect of the spectrum flow. Here a researcher doesn't face the issue of the bandwidth selection (as in Short Time Fourier Transform). Wavelet transform allows data flow contraction, which characterizes the mode of the electric power system. This method allows determining effective value of current and voltage during a nonsinusoidal mode.

Basing on the received wavelet coefficients after transformation we can receive active and distortion power for individual harmonic, which allows précising technical and economic calculations when choosing measures for harmonic filtering.

References

[1] Gavrilin A N, Bolovin E V, Glazyrin A S, Kladiev S N and Polishchuk V I 2019 Resonant oscillations with a limiting amplitude in a vibration electromagnetic activator *Bulletin of the Tomsk Polytechnic University. Geo assets engineering* 2019 vol 330 **1** pp 201–13

[2] Gavrilin A N, Kladiev S N, Glazyrin A S, Bolovin E V and Polishchuk V I 2019 Identification of parameters of vibration electromagnetic activator mechanical system using limiting nearresonance frequency *Bulletin of the Tomsk Polytechnic University. Geo assets engineering* 2019 vol 330 **4** pp 158–77

[3] Glazyrin A S, Anikin V V, Bunkov D S, Antyaskin D I, Startseva Y N, Kovalev V Z, Khamitov R N, Kladiev S N and Filipas A A 2020 Nonlinear algebraic estimation of a vibration electromagnetic activator inductivity by a failing current curve *Bulletin of the Tomsk Polytechnic University, Geo Assets Engineering* 202, vol 331 **1** pp 148-57

[4] Qian C A 2018 Power waveform classification method for adaptive synchrophasor estimation *IEEE Transactions on Instrumentation and Measurement* vol 67 **7** pp 1646-58

[5] Chicco G, Pons E, Russo A, Spertino F, Porumb R, Postolache P and Toader C 2017 Assessment of unbalance and distortion components in three-phase systems with harmonics and interharmonics *Electric power systems research* 2017 vol 147 pp 201-12

[6] Czarnecki L S and Haley P M 2016 Power properties of four-wire systems at nonsinusoidal supply voltage *IEEE Transactions on Power Delivery* 2016 vol 31 **2** pp 513-21

[7] Silva S, Costa P, Gouvea M, Lacerda A, Alves F and Leite D 2018 High impedance fault detection in power distribution systems using wavelet transform and evolving neural network *Electric power systems research* 2018 vol 154 pp 474-83

[8] Monteiro H L M, Duque C A, Silva L R M, Meyer j, Stiegler R, Testa A and Ribeiro P F 2017 Harmonic impedance measurement based on short time current injections *Electric power systems research* 2017 vol 148 pp. 108-16

[9] Gil-de-Castro A, Rönnberg S K and Bollen M H J 2017 Light intensity variation (flicker) and harmonic emission related to LED lamps *Electric power systems research* 2017 vol 146 pp 107-14

[10] Barros J, Gutierrez J J, De Apraiz M, Saiz P, Diego R I and Lazkano A 2016 Rapid voltage changes in power system networks and their effect on flicker *IEEE Transactions on Power Delivery* 2016 vol 31 issue 1 pp 262-70

[11] Osipov D S, Goryunov V N, Faifer L A, Kisselyov B Y and Dolgikh N N 2017 Development of conductive parts power losses calculation method in case of interharmonics *Przegląd Elektrotechniczny* 2017 NR 6 pp 146-149

IOP Conf. Series: Materials Science and Engineering 1118 (2021) 012032 doi:10.1088/1757-899X/1118/1/012032

The possibility to use Bandsaw Dynamic Tensioning in Bandsaw Machines

V K Shilko, V A Slepchenko and I V Slepchenko

National Research Tomsk State University, 36, Lenina ave., Tomsk, 634050, Russia

E-mail: docent46@yandex.ru

Abstract. The authors considered the mechanism of motion transmission to wood-cutting band saws, using the effects of their dynamic tension during operation, within the general statements of the flexible link transmission theory.

1. Introduction

Bandsaw machines are widely used in the woodworking, light, food, and metalworking industries. Basic working principles of the cutting mechanisms in the bandsaw wood-cutting machines, the ways of their development, and physical processes during the main motion pulsion are considered in the linked articles [1-7].

One of the development possibilities of the cutting mechanisms in the bandsaw wood-cutting machines appeared quite recently (about 20-30 years ago). It is based on the wood locally stretched in the cutting area by a band saw and is worth viewing as it moves many tensions in the bandsaw away. You can see a typical scheme of this machine in Figure 1, [8]. The authors refuse a double-pulley cutting mechanism scheme and move to the scheme, where the mechanical frictional contacts transmit the motion.

Saw tension appears because of the difference in effort of the saw transmission and braking forces. This way of tensioning in a bandsaw is called dynamic tensioning. The main problem of this method of machine development is the unpredictability of the saw out of the cutting area when the cutting speed is high (30…35 m/s). Besides, the friction contact, as noted earlier [1-3], is unreliable and causes increased wear of the band saw.

2. Analytical review

For a more stable transmission of movement to the band saw, it is necessary to introduce several friction pairs into such schemes (Figure 1) that transmit motion (and torsion), or use traction elements based on tracks, or belts, as shown in the works [9-12]. Besides, in such systems, as in all non-pulley schemes of band saws, a system for adjusting the saw position and keeping it from shifting under the influence of the regular cutting force is not fully developed. The proposed solutions for holding the saw on friction support with the help of thrust rear rollers, as the development history of band-saw machines shows, do not reduce the existing problems.

Decreasing the essential stresses in band saws, this direction of cutting mechanisms development does not solve other problems.

Content from this work may be used under the terms of the Creative Commons Attribution 3.0 licence. Any further distribution of this work must maintain attribution to the author(s) and the title of the work, journal citation and DOI.
Published under licence by IOP Publishing Ltd

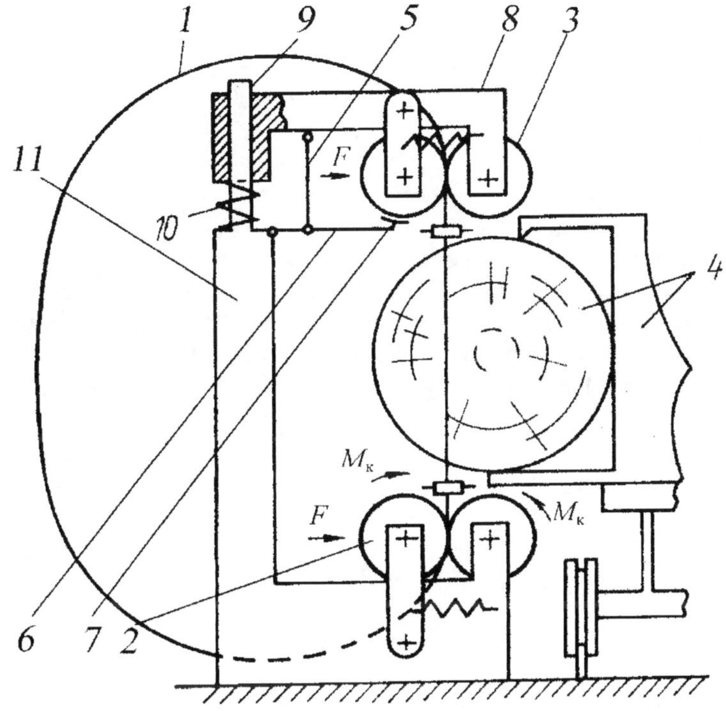

Figure 1. A non-pulley bandsaw machine scheme with bandsaw tention only in the cutting area [8]: 1 – saw, 2, 3 – drive and brake rollers, 4 – log feeding and fastening system, 5, 6, 7 – elements of the friction brake, 8, 9, 10, 11 – frame with brake guide device.

They are related to the development of efficient support circuits for the saw outside the cutting area, as well as to providing stable support reactions for the band saw near the cutting area to counteract the feed force. In this regard, this direction of development of band saws with a locally stretched band saw only in the cutting area has not yet received practical application in the woodworking industry. Some analogues of such machines are used in light industry for cutting leather and fabrics and in the metalworking industry as cutting machines. At the same time, the cutting speed is 2 times less (10...15 m/min) than when sawing wood. Therefore, the saw outside the cutting area does not suffer from additional dynamic effects.

Despite many unresolved issues, these schemes help to develop theoretical models and laboratory samples for analytical and experimental research in the field of band sawing wood.

3. The concept model of developement

The concept model of the band saw cutting mechanism must meet the following requirements:

– have tension for the band saw only in the cutting area between the guides,

– do not have additional centrifugal forces in the working branch of the saw,

– have reliable adjustments to give the band saw the desired position,

– create stable support reactions for the saw near the cutting zone to counteract the feed force based on relative rest friction.

In this case, the band saw will only be stretched in the cutting area due to the required pulling force F_{pul} of the drive-guide device, which should be

$$F_{pul} \geq (F_{br} + P_z)C_t \quad (1)$$

where P_z is a tangent component of cutting forces;

IOP Conf. Series: Materials Science and Engineering 1118 (2021) 012032 doi:10.1088/1757-899X/1118/1/012032

F_{br} is a required braking force of the band saw;

$C_t \approx 2,5...3$ is the coefficient of tractive effort reserve.

The effort F_p, required for tension of the band saw in the work area, will be as follows:

$$F_p \geq F_{pul} - F_{br} \geq 2F_0 - F_x \tag{2}$$

where F_0 stands for the pre-tensioning force of the band saw branch in static;

F_x is tension force in the idle branch of the band saw during operation.

In general, the implementation of these conceptual schemes of guide devices and cutting mechanisms entails a complication of the existing band saw equipment [1] and is not yet fully justified, since little has been studied theoretically and practically, especially in terms of maintaining the band saw outside the cut. Therefore, it is necessary to limit yourself to simpler schemes that do not complicate band saws, while at the same time significantly increase the stability of band saws.

4. The main result

From a practical point of view, double-pulley systems with dynamic tension of a pre - "weakly stretched" saw (Figure 2) and a locally stretched section of it in the cutting area can be more efficient due to the Poncelet ratio [1]. At the same time, it is necessary to solve the issue of transmitting the traction force by a two-way drive-guide device and partially braking the saw movement by a brake-guide device.

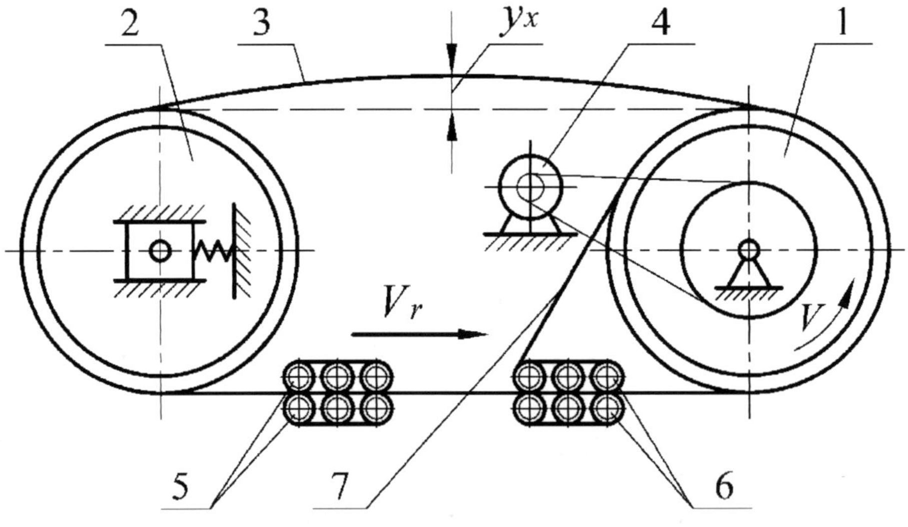

Figure 2. Layout diagram of a horizontal band saw with dynamic tension of a pre-stretched band saw: 1, 2 – master and slave pulleys, 3 – a band saw, 4 – electric drive motor, 5, 6 – brake and drive guide devices, 7 – a traction flexible intermediate working body, yx – boom deflec-tion of the idle branch of the band saw when working from the redistribution of the tension force in accordance with the Poncelet ratio.

In comparison with this scheme, the cutting mechanism of a traditional two-blade band saw with rubberized rollers placed on the driven and driving pulleys will be simpler (Figure 3). This will allow for a lower pre-tensioning force of the saw to increase the pulling force and achieve the effect of dynamic tension due to its weakening in the idle branch.

The effect of dynamic tension of the band saw is present in any conventional cutting mechanism scheme of a double-pulley band saw and is determined from the Poncelet ratio.

$$F_0 = \frac{F_p + F_x}{2} \tag{3}$$

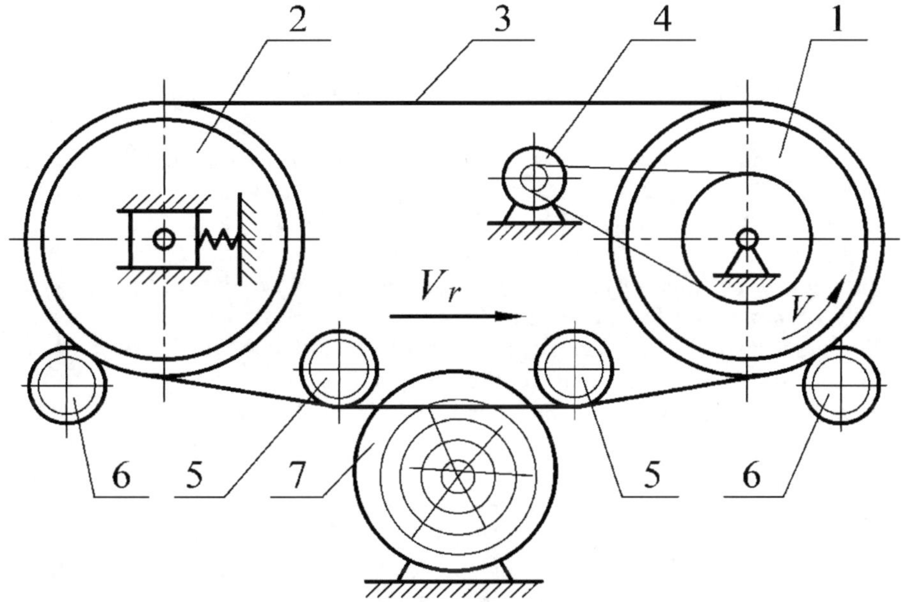

3)

Figure. 3. Recommended installation scheme for rubberized guide rollers in the cutting mechanisms of double-pulley band saws with the effect of the band saw dynamic tension. 1, 2 – master and slave pulleys; 3 – a band saw; 4 – electric drive motor; 5, 6 – guiding devices; 7 – sawn timber.

However, given that Fp >> Fx due to the resistance to the saw movement from the pulleys and guides, as well as due to the action of cutting forces, for the full force compensation, the tension in the working branch of band saw blades should be subject to the following conditions (2)

$$F_{pul} - (F_{br} + P_z)c_t \geq 2F_0 - F_x \tag{4}$$

Thus, based on the theoretical studies of the conditions for transmitting movement to the band saw, there is a need to create a new direction in the development of cutting mechanisms for band saws. The direction should involve the establishment of reliable support reactions for the band saw in the cutting area, preventing its unauthorized transverse movement along the pulleys under the action of the feed force. One can achieve it by using rubberized rollers in the cutting mechanism of band saws to get the effect of the dynamic tension of the band saw.

This direction of cutting mechanisms development for woodworking band saw machinery allows you to combine the advantages of a traditional machine with a new way of their evolution. That is, to create double-pulley band saws with dynamic tension of a pre - "weakly stretched" band saw in the case of using some drive and brake guide devices following conditions (2) and (4).

The tangible advantages of this direction of band-saws development include the fact that they maintain continuity with traditional double-pulley schemes. This fact makes it possible to use technical engineering not only for creating new band saws but also for upgrading existing ones and those in operation.

5. Conclusions

1. One of the promising directions for the development of cutting mechanisms for band saws is the direction based on sawn wood locally stretched in the cutting area with a band saw using the effect of dynamic tension of the band saw.

2. To achieve the effect of the dynamic tension of the band saw, one can use rubberized pres-sure rollers in the systems of cutting mechanisms of traditional double-pulley band saws.

3. The tangible advantages of this direction of cutting mechanisms development for band saws are that they maintain continuity with traditional double-pulley schemes.

References

[1] Shilko V K 2005 Cutting Mechanisms of Band saw Machines *Tomsk: TSUAB Press* p 220
[2] Kondratyuk A A, Shilko V K and Rudnev V D 2004 Estimation of band mill traction per formances *8th International Simposium on Science and Technology KORUS* vol **1** pp 24–27
[3] Shilko V K 2004 Prospects of bandsaw development *Woodworking industry* vol **5** pp 6–11
[4] Druzhkov G F 1983 Band saws for sawing wood *M:Forest industry* p 72
[5] Kondratyuk A A, Shilko V K and Novoseltseva I V 2004 Modern trends in the development of cutting mechanisms for band saws *The Int. Conf. on Mechanical Engineering and Modern Tech (Tomsk: TPU Press)* pp 314–315
[6] Shilko V K 2004 Use of the friction of relative rest in cutting mechanisms of belt sawing machines. *Friction and Wear* vol **25** № 5 pp 512–518
[7] Rudnev V D, Kondratyuk A A and Shilko V K 2004 Band saw machines on the basis of long contact friction relative peace *Current problems of the forest complex Proc Int Conf 5 NMRC* “Les 2004” (Bryansk) pp 244–247
[8] A M Lopatin A M, Dolgopolova N A 2001 A Bandsaw Machine Patent 2162404 RF IPC 7B 27 B15/04 *Inventions Useful Models* vol **3** p 312
[9] Slepchenko I V, Kondratyuk A A and Shilko V K 2005 Influence of dynamic factors on the stability of band saws Modern Technic and Technologies *Proc. Int. Conf. XI NMRC Modern Technic and Technologies* Tomsk: TPU vol **1** pp 167–169
[10] Vlasov V P, Shilko V K, Slepchenko I V and Kondratjev M JU 2005 A Bandsaw Machine pat 45672 RF IPC 7B 27 B 15/00 13/00 *Inventions. Useful Models* vol **15** Part V p 972
[11] Shilko V K, Kondratjev M JU, Slepchenko I V and Stjepanov D E 2007 A Bandsaw Machine pat 64555 RF IPC 7B 27 B 13/00 15/00 15/04 *Inventions. Useful Models* vol **1** part II p 519
[12] Shilko V K, Kondratjev M JU, Slepchenko I V and Stjepanov D E 2008 A Bandsaw Machine pat 70480 RF IPC 7B 27 B 13/00 15/00 15/04 *Inventions. Useful Models* vol **3** part 5 p 1222

Influence of some parameters of magnetization and cutting modes on the tool life when machining construction materials

S M Khasanov and A A Yakubov
Mechanics of Materials Department, Tashkent State Technical University, 100093, University Street, Tashkent city, Uzbekistan

E-mail: asroryakubov.1974@mail.ru

Abstract. The article examines the study of the influence of the magnetic field on the wear resistance of tool materials. The issues of optimal modes of magnetization and cutting modes, as well as the influence of the rake angle are considered.

1. Introduction

Many different methods of increasing wear resistance of the cutting tool are known. They are nitriding, cyanation, processing by cold, cutting or processing of the tool in magnetic field, application of different lubricant cooling mediums, etc. Recently use of magnetic field in machining process was conducted in the different directions [11].

In [1], the possibilities of using a pulsed magnetic field to improve the complex of mechanical properties of copper and alloys D16, BrB2 and L63 are considered. It is shown that the effect of a pulsed magnetic field increases the characteristics of plasticity, fatigue strength and tribotechnical parameters.

A team of authors [2] carried out extensive work on the use of magnetic fluids as technological media in the machining of metals. They magnetized the tools directly during processing. The study was carried out in drilling and milling with magnetic fluids. The results showed an almost twofold increase in the tool life of drills and mills.

In work [3, 4], it is checked the effect of Herbert in quick cutting steels. Results of cross-validation testings of firmness of cutting tools are given in an initial status and after magnetization and also magnetic and abrasive processing is given. According to them processing of details in magnetic field ferromagnetic powders is the integrated method which is combined mechanical effect with use of magnetic field energy. As a result, Hardening of a surface layer was noticed.

The authors in [5] consider the use of magnetic microcapsules as LCTM (Lubricating and cooling technological means). They found that the use of LCTM microcapsules with a concentration of microcapsules (CM) from 1.0 to 4% promotes an effective increase in the durability of cutting tools not only in comparison with dry cutting, but also when using standard water-based and oil-based cutting tools.

Authors Skirdenko O.I. et al. [6] studied the state of the art of processing tool materials in magnetic and magnetic-pulse fields.

In their works, Ben A.P. and others [7] gave a device for magnetic-pulse processing of metal products with an automated selection of processing modes.

Content from this work may be used under the terms of the Creative Commons Attribution 3.0 licence. Any further distribution of this work must maintain attribution to the author(s) and the title of the work, journal citation and DOI.
Published under licence by IOP Publishing Ltd

One of the methods for increasing the wear resistance of cutting tools is the magnetization of the cutting tool [8, 9, 10].
Many works have been devoted to the question of the influence of the magnetization of high-speed tools on the productivity of machining and on the wear resistance of cutters.
Numerous experiments have established a significant effect of the magnetic field on the durability of the cutting tool. However, the optimal modes of magnetization have not yet been determined.
The analysis of literary data shows that the effect of using magnetic field in the course of cutting depends on such factors as time of magnetization and magnetic field strength. The strength of the magnetic field is directly proportional to the strength of the current in the circuit. Therefore, to determine the optimal value of the magnetization time - and the current strength - J, we determined the dependences $T = f(\tau)$ and $T = f(J)$ for non-magnetized and magnetized high-speed cutters when processing steel 25. Then we determined the coefficient of increasing the durability of the magnetized tool by the following formula:

$$K_T = \frac{T_{dmt}}{T_{dnmt}} \quad (1)$$

where: T_{dmt} - durability of a magnetized tool;

T_{dnmt} - durability of non-magnetized tool.

2. Experiments and results

The experiments were carried out at cutting modes v= 120 m / min; S = 0.11 mm / rev; and $\gamma = 15^0$, at which a stable effect of magnetization was observed. Figure 1 and Figure 2 show the influence of the magnetization time and current strength on the efficiency factor of the magnetic field, from which it can be seen that the optimal value of the magnetization time is τ= 3 minutes, and the current strength is J = 3.75 amperes. Similar data were obtained when processing steels 40X and 45. Therefore, in all subsequent experiments, the magnetization of the cutting tool was carried out at τ=3 min, and J = 3.75 amperes.

It is clear that the strength of the magnetic field created by the coil at constant values of the magnetization modes depends on the size of the tool, which plays the role of the core. Therefore, a change in the length of the tool as a result of its regrinding leads to a change in the strength of the magnetic field created by the coil, and therefore to a change in the remanent magnetization.

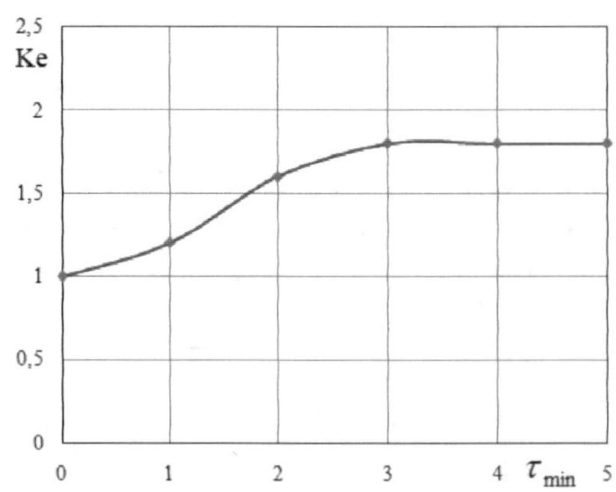

Figure 1. Influence of the magnetization time on the coefficient of life of the cutting tool when processing steel 25

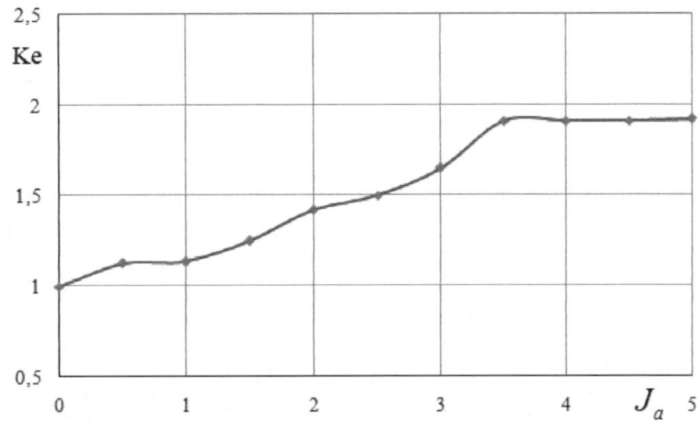

Figure 2. The influence of the current on the coefficient of life of the cutting tool when processing steel 25.

In order to check the influence of this factor on the wear resistance of high-speed tools, experiments were carried out to determine the residual magnetic induction with cutters having different lengths.

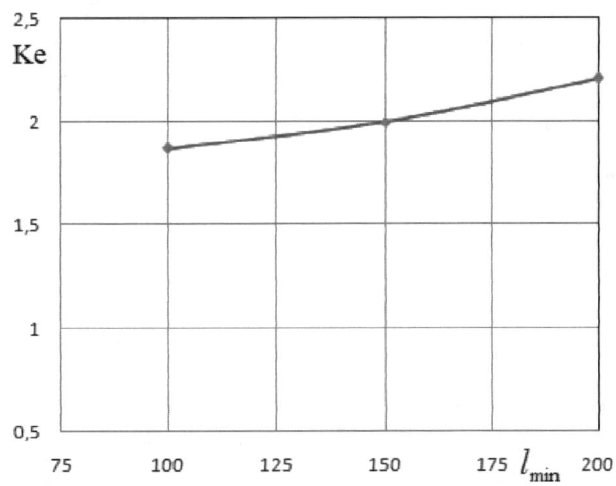

Figure 3. Influence of the length of the cutting tool on the coefficient of life of the cutting tool when machining steel 25.

The measurement of the residual magnetic induction was carried out by the ballistic method. For 3 cutters made of high-speed steel P18, having a length of 100; 150; 200 mm, preliminary resistance experiments were carried out when processing steel 25 at $S = 0.11$ mm/rev; $v = 152$ m/min. After that, these cutters were magnetized at the same modes of magnetization ($J = 3.75$; $\tau = 3$ min), and then the residual magnetic induction was measured. The results obtained showed that at $l_1 = 100$ mm $Br_1 = 0.33334$ tesla, at $l_2 = 150$ mm tesla, at $Br_3 = 0.36877$ mm tesla.

In order to determine the influence of the value of the residual magnetic induction on the wear resistance of the tool, resistance experiments were carried out for the above magnetized cutters. The data obtained showed (Fig. 3) that a change in the length of the cutters under the conditions of the experiments carried out does not significantly affect the effect of the use of magnetized tools.

One of the factors influencing the cutting temperature and plastic deformation of the cut layer is the rake angle γ. If the above assumptions are correct, then the effect of the magnetization of the cutting tool in a quantitative ratio should depend on the value of the rake angle γ. To check this, experiments were carried out on steel 25 with a P18 high speed cutter at feed rates $S = 0.11; 0.15; 0.2$ mm / rev. In these experiments, the dependence T = f (γ) was initially recorded for a non-magnetized cutter, and then the same cutter was magnetized and the corresponding dependence for a magnetized cutter was recorded.

Experiments have shown that in a wide range of values of the rake angle γ, (0 - 20^0), the magnetization of the cutter in comparison with the non-magnetized one increases its durability. However, the effect of tool magnetization changes quantitatively with changing rake angle γ. The optimal rake angle γ for magnetized tools for all investigated feeds is $\gamma = 15^0$. This can be explained by the optimal combination of plastic deformation of the cut layer and cutting temperature, contact surface area, heat removal conditions, and the value of the friction coefficient.

The next series of experiments was carried out at a constant value of the angle $\gamma = 15^0$ for various cutting speeds and feeds when machining steels 25 and 45. The results of the experiments are presented in Figures 4, 5, 6, 7. From the given data it can be seen that in the entire range of cutting speeds, in which the experiments were carried out, there is an increase in the resistance of the magnetized cutting tool.

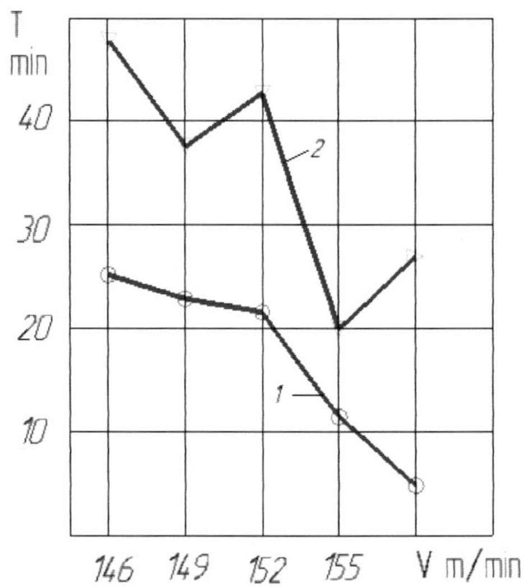

Figure 4. Influence of cutting speed and cutting tool magnetization on the tool life when machining steel 25, S = 0.11 mm / rev
1. Non-magnetized tool
2. Magnetized tool

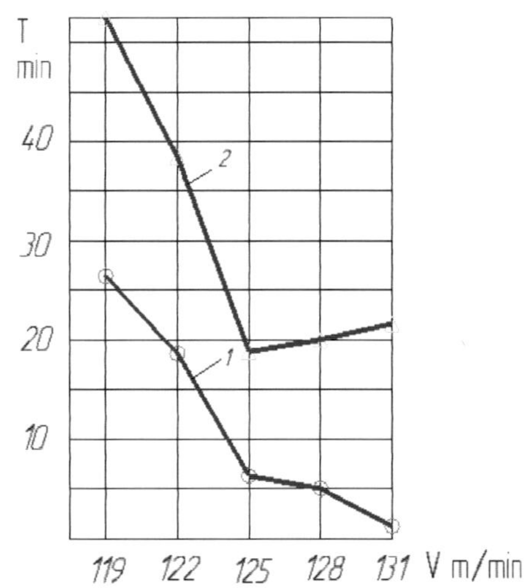

Figure 5. The effect of the cutting speed and the magnetization of the cutting tool on the durability of the cutting tool when processing steel 25, S = 0.15 mm / rev
1. Non-magnetized tool
2. Magnetized tool

However, there is a clear tendency towards a decrease in the effect of magnetization with increasing feed. This can be explained by the fact that thin chips heats up and deforms better and therefore their shrinkage is higher compared to thick chips due to an increase in the coefficient of friction. The results of friction experiments also showed that the greatest decrease in the coefficient of friction due to magnetization under experimental conditions occurs at low specific pressures and friction rates.

Hence, we can conclude that at lower values of feed, the decrease in the coefficient of friction due to magnetization of the tool material occurs by a large amount, and therefore, with a decrease in the feed value, the efficiency of magnetized tools increases.

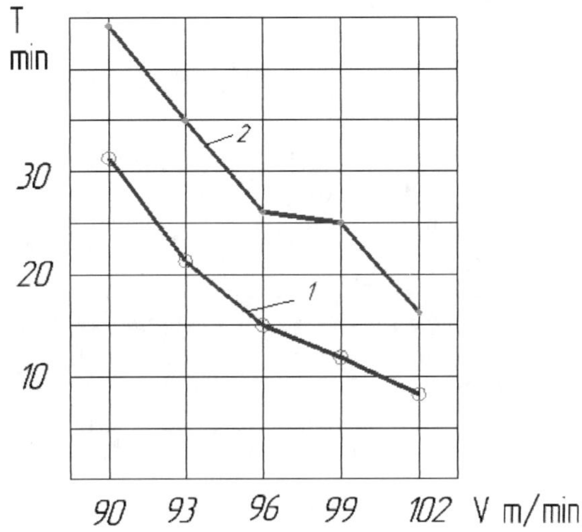

Figure 6. The effect of cutting speed and magnetization of the cutter on the durability of the cutting tool when processing steel 25, S = 0.2 mm / rev
1. Non-magnetized tool
2. Magnetized tool

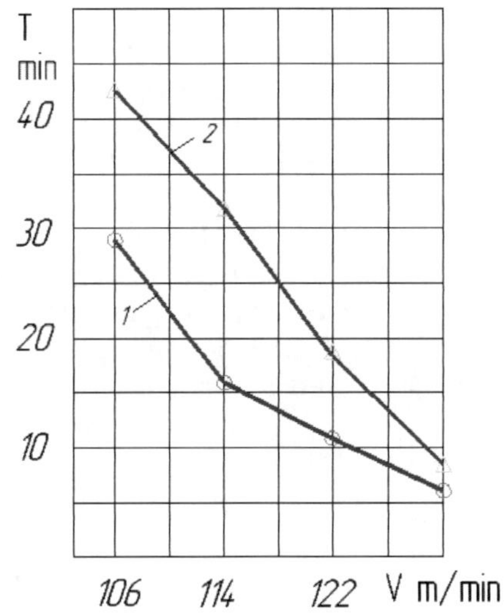

Figure 7. Effect of cutting speed and cutter magnetization on the tool life when machining steel 45, S = 0.11 mm / rev, γ = 150 P6M5
1. Non-magnetized tool
2. Magnetized tool

2. Conclusions

Based on the performed theoretical studies and analysis of the obtained experimental data, the following conclusions can be drawn:

1. The optimal modes of magnetization of high-speed tools for their effective use are $J_{onm} = 3.75$ ampere, $\tau_{onm} = 3$ min;
2. Residual magnetic induction with a change in the length of the cutting tool changes slightly and does not significantly influence the effect of the use of magnetized tools.
3. Rake angle $\gamma = 15^0$ and cutting speed when machining structural carbon steels have a significant effect on the efficiency of magnetized tools and this effect decreases with increasing feed.

References

[1] Zdor G N, Anisovich A G and Yaskovich A G 2004 Application of a pulsed magnetic field to improve the mechanical properties of non-ferrous metal alloys *Problems of Mechanical Engineering and Reliability of Machines* **5** [in Russian]

[2] Orlov D V et al. Application of magnetic fluids as technological media in mechanical processing of metals, http://magneticliquid.narod.ru/autority/093.htm

[3] Dobroskok V L et al 2018 Classification of integrated machining processes for materials». *Cutting and tools in technological systems*, **83** [in Russian]

[4] Pustovoyt V N and Karnilov Yu A 2007 Investigation of the Gebert effect for increasing the tool life of cutting tools. *Strengthening Technologies and Coatings* **13** 55-61

[5] Naumov A G and Latishev V G 2011 Application of magnetic microcapsules as cutting fluid for metal cutting. *Bulletin of the Ivanovo State Power Engineering University* **2**

[6] Skirdenko O I, Klevsov K M, Sharko A V and Mikhaylik V D 2011 State of the art of processing tool materials in magnetic and magnetic-pulse fields. *Scientific Bulletin of the Kherson State Maritime Academy* **2**
[7] Ben A P, Malgin B V and Vrubevskiy R Ye 2911 Device for magnetic-pulse processing of metal products with automated selection of processing modes. *Scientific Bulletin of the Kherson State Maritime Academy* **2**.
[8] Kindenko N I 2011 Classification of methods for magnetic processing of cutting tools. *Scientific Bulletin of the Donbass State Machine-Building Academy* **2 (8E)** 118-121
[9] Khasanov S M 2016 Some questions of the effectiveness of the use of a magnetic field in mechanical engineering. *Proceedings of Modern materials, equipment and technology in mechanical engineering, 19-21 April, Andijan, Uzbekistan*
[10] Khasanov S M and Umarov E O 2015 *The influence of the magnetization of a high-speed cutter on its durability when processing paramagnetic materials* (Rostov: Moscow State Technical University of Civil Aviation, Rostov Branch)
[11] Umarov E O, Mardonov U T and Shoazimova U Kh 2020 Influence of the Magnetic Field on the Viscosity Coefficient of Lubricoolant that is used in the Cutting Proces, *International Journal of Mechatronics and Applied Mechanics* **8 2** 144-149

Flexural properties of copper-filled carbon fiber reinforced polymers

Eremin A V[1,2] and Burkov A V[1,2]

[1]Institute of Strength Physics and Materials Science, Akademicheskiy avenue 2/4, Tomsk, 634055, Russia
[2]Tomsk Polytechnic University, Lenin avenue 30, Tomsk, 634050, Russia

Corresponding author: burkovispms@mail.ru

Abstract. The paper demonstrates the results of flexural tests of carbon fiber reinforced polymers with the addition of copper microfillers with content varying from 2.5 to 10 wt%. It is shown that the implemented dispersive reinforcement leads to the increase of flexural strength on composites by 9%, while the elastic modulus remains the same. The reasons for the advancement in mechanical properties are discussed.

1. Introduction

Composite materials become a very important and widely used material in various fields, including but not limited to aerospace, automotive, marine, sport equipment, etc. Structural application of composites is connected to the possibility of adjusting their physical and mechanical properties to the particular area and operational conditions. Nowadays, special attention is paid for the polymers and one of the composites combining high specific strength and good fatigue resistance – carbon fiber reinforced polymers (CFRP). CFRPs due to their superior mechanical properties find their niche in aircrafts structural elements as fuselage, wings and tail. However, the application is limited due to sensitivity to impact damage and delamination. One of the possible ways to increase the desired properties is additional reinforcement of the laminated composite by dispersive nano- or microparticles. In case of aircraft industry, it not only could provide advancement in mechanical characteristics, but potentially could increase the important functional features such as electrical and thermal conductivity.

Addition of carbon fillers are widely reported in different papers and contain some information on mechanical and physical properties. For example, in the published paper [1] it has been shown that composite reinforced by dispersive fillers has higher mechanical properties under tension, compression, bending and fatigue. In the few other articles [2,3], the increase of fracture strength under impact loading has been reported.

The results on metals fillers are less studied – in the following papers the composites were filled with silver [4-5], copper metal spheres [6] and metal coatings [7], which significantly improve functional properties. In [4], silver is used as a coating of carbon fibers which then reinforce the epoxy binder leading to an increase in electrical conductivity by 4 times. In [5], the addition of silver powder increased the thermal conductivity of polypropylene composites by 4 times with 5 vol% powder content and also reduced the coefficient of thermal expansion linearly.

In [7], four coating methods were investigated: (1) hybrid coating based on silver nanoparticles dispersed in a conducting polymer PEDOT: PSS; (2) hybrid coating in the form of silver-coated nanotubes, which was applied to the surface of the composite; (3) A solid metallic coating of the

Content from this work may be used under the terms of the Creative Commons Attribution 3.0 licence. Any further distribution of this work must maintain attribution to the author(s) and the title of the work, journal citation and DOI.
Published under licence by IOP Publishing Ltd

composite with silver and (4) a metallic coating of tin with 10% copper. To test the materials obtained, damage to the substrate during the course of electrical discharges (similar to a lightning strike) was evaluated. Conventional carbon and carbon plastic with foamed copper foil on the surface were taken as base materials. According to the test results, it was found that the foamed copper foil showed the highest protective ability, and the metal coatings (3) and (4) exceeded the hybrid ones (1) and (2).

The most papers describe the effect of metals addition to the electrical properties, while only limited amount of them is devoted to mechanical characteristics. The present paper study the flexural properties of CFRP modified by copper micropowder.

2. Materials and testing technique

The specimens were carbon fiber reinforced polymers (CFRP) with addition of copper microparticles. The lay-up was balanced, symmetric and pseudosotropic: $[0/90;+45/-45]_{4S}$. CFRP components:

- CBX300 biaxial fabrics made of PAN carbon fiber (Mitsubishi Pyrofil TR50S 12K);
- R&G Epoxy L with GL2 hardener;
- copper micropowder with a mean diameter of 50-100 μm.

The resulting thickness of the plates was ~4 mm and the specimens were cut using milling machine. The content of the fillers was varied resulting in the formation of 4 concentration of uCu in the blanks. The compositions tested are summarized in Table 1. The additive concentration is calculated for the total epoxy weight, not for the resulting CFRP.

Two types of 3-point bending tests were utilized in the paper. The first was done according to ASTM D2344 which is aimed at the evaluation of interlaminar shear strength by flexural loading of short-beam specimen. The size of the specimen was 24×8×4 mm and a span length was 16 mm. The second type 3-point bending test was arranged according to ASTM D7264. Five specimens having the sizes of 154×13×4 mm were tested for each composition. The span length of 128 mm resulted in a span-to-thickness ratio of 32:1. The loading rate was equal to 5 mm/min. Flexural strength and modulus were measured after the tests.

The resulted CFRP plates and long-beam bending specimens are presented in Figure 1. The addition of copper slightly changes the color of the originally black epoxy emerging light ochre tint. The higher uCu content the more saturated the tint.

Table 1. Overview of the tested compositions

Additive	**Content, wt%**	**Designation**
Non-modified CFRP	0	NM CFRP
uCu	2.5	uCu-2.5
	5	uCu-5.0
	7.5	uCu-7.5
	10	uCu-10

3. Experimental results and discussion

The summary of the results on short-beam flexural tests are demonstrated in Table 2. This test type do not implies the determination of the elastic modulus or strain at break, so the main characteristic being obtained is the interlaminar shear strength. Non-modified epoxy has strength of 50.7 MPa which is common for such composites. The addition of 2.5 and 5 wt% of micro copper powder slightly increases the property up to 51.9 MPa and 51.8 MPa – the percentage growth is +2.5% and +2.3% respectively. The improvement of the interlaminar strength for low copper content is mostly connected with scatter rather than with any physical processes affecting the results. While 7.5 wt% gives some potential and demonstrates the advancement in mechanical characteristic to 55.8 MPa (+8.5%). However, the further increase of the content leads to the reduction of the strength (53.7 MPa and +5.9% compared to the NM CFRP).

a b

Figure 1. Images of the uCu-7.5 blank (a) and uCu-100 specimens for ASTM D7264 testing (b)

Of course, the changes are minor but they obviously take place. The authors attribute the increase in the interlaminar properties to the addition of the cooper micropowder. The micro particles are involved in the epoxy deformation process as obstacles which strength the matrix and increase its properties. At the same time the adhesion between epoxy and copper could not be high enough because it is mostly mechanical, so it could not provide the great increase and after some additive concentration the lack of chemical adhesion will become a major effect which results in decrease of the mechanical characteristics.

Table 2. Results of short-beam bending test

Material	**Interlaminar shear strength, MPa**
NM CFRP	50.7±4.1
uCu-2.5	51.9±3.6 (**+2.5%**)
uCu-5	51.8±3.6 (**+2.3%**)
uCu-7.5	55.8±3.6 (**+8.5%**)
uCu-10	53.7±2.8 (**+5.9%**)

The results of for the second type of bending test – long-beam flexural – are presented in the form of plot in figure 2. At the upper part of the diagram the changes in elastic modulus of CFRP versus uCu content is demonstrated, the lower part is for combined graph on flexural strength and strain at break.

The elastic modulus of non-modified CFRP is 53.7 GPa and addition of copper micropowder do not change it so much. The moduli have values of 54.3, 53.1, 52.5 and 53.0 GPa for 2.5, 5, 7.5 and 10 wt% of micro Cu. So the elastic modulus is not affected by the addition of copper powder in long-beam flexural tests. However, the strength and elongation at break were increased. The first changed from 650 MPa to 710 MPa, which is approximately 9% of the increase. Subsequently, the strain at break become higher: simply the specimen could withstand higher deformation due to higher strength while the modulus is the same.

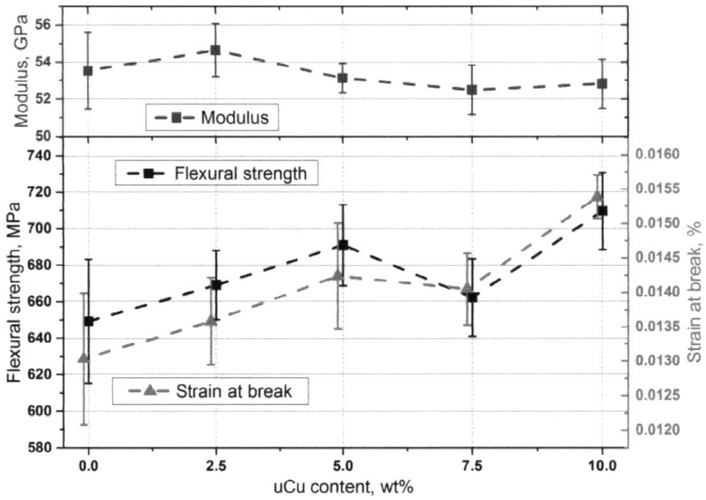

Figure 2. Flexural properties obtained in long-beam bending test

4. Conclusion

The experimental results obtained on the effect of copper micropowder addition into epoxy-based CFRP lead to the following conclusions:

- Copper microfiller increase the strength of the CFRP in case of short-beam and long-beam flexural tests – the approximate enhance reach 9% for both test types.
- The flexural modulus remains the same and it is not affected by the addition of copper powder.
- The assumed reason for the observed effects is the analogous to doping strengthening in metals when the copper particles play a role of obstacles to the development of failure at low content. For higher amount of copper due to the lack of chemical adhesion between powder and epoxy the filler become a concentrator for the crack initiation and mechanical properties reduce.

Acknowledgements

This work has been supported by the Russian Scientific Foundation, grant 19-79-10148

References

[1] Knoll J B, Riecken B T, Kosmann N, Chandrasekaran S, Schulte K and Fiedler B 2014 The effect of carbon nanoparticles on the fatigue performance of carbon fibre reinforced epoxy *Compos. Part A Appl. Sci. Manuf.* **67** 233–40

[2] Quan D, Urdániz J L and Ivanković A 2018 Enhancing mode-I and mode-II fracture toughness of epoxy and carbon fibre reinforced epoxy composites using multi-walled carbon nanotubes *Mater. Des.* **143** 81–92

[3] Tehrani M, Safdari M, Boroujeni A Y, Razavi Z, Case S W, Dahmen K, Garmestani H and Al-Haik M S 2013 Hybrid carbon fiber/carbon nanotube composites for structural damping applications *Nanotechnology* **24** 155704

[4] Ha M-S, Kwon O-Y and Choi H-S 2010 Improved Electrical Conductivity of CFRP by Conductive Nano-Particles Coating for lightning Strike Protection *J. Korean Soc. Compos. Mater.* **23** 31–6

[5] Maiti S N and Ghosh K 1994 Thermal characteristics of silver powder-filled polypropylene composites *J. Appl. Polym. Sci.* **52** 1091–103

[6] Yu G-C, Wu L-Z, Feng L-J and Yang W 2016 Thermal and mechanical properties of carbon fiber polymer-matrix composites with a 3D thermal conductive pathway *Compos. Struct.* **149** 213–9

[7] Rajesh P S M, Sirois F and Therriault D 2018 Damage response of composites coated with

conducting materials subjected to emulated lightning strikes *Mater. Des.* **139** 45–55

IOP Conf. Series: Materials Science and Engineering 1118 (2021) 012035 doi:10.1088/1757-899X/1118/1/012035

Thermogravimetric analysis of epoxy-based carbon fiber reinforced polymers modified by carbon fillers

Burkov M V[1,2] and Eremin A V[1,2]

[1]Institute of Strength Physics and Materials Science, Akademicheskiy avenue 2/4, Tomsk, 634055, Russia
[2]Tomsk Polytechnic University, Lenin avenue 30, Tomsk, 634050, Russia

Corresponding author: burkovispms@mail.ru

Abstract. The paper presents the results of thermogravimetric/differential scanning calorimetry investigation of hybrid thermoset CFRPs modified by the addition of single-wall carbon nanotubes and milled carbon fibers. Specimens were tested using NETZSCH STA 409 PC/PG with the following parameters: heating from ambient temperature to 873 K with a rate of 15 K/min; Ar flow rate of 50 ml/min. The content of the carbon filler in CFRP was varied allowing to establish some regularities in thermal and physical properties.

1. Introduction

Composite materials are applied in many fields such as aerospace, marine, and mechanical engineering. Carbon fiber reinforced polymers (CFRPs) are one of the most versatile materials among composites with polymer matrices. They have outstanding properties: high specific strength, good fatigue resistance, etc. One of the novel ways to improve mechanical and physical properties is associated with the addition of nano- or micropowders. Thus hybrid composites where carbon continuous fibers or fabric is additionally reinforced by such dispersive fillers are obtained. It has been shown that hybrid composites demonstrate higher mechanical properties under tension, compression, bending and fatigue [1]; impact properties are investigated in [2-4].

In [2] Dong Quan et al reported about the improvement of the fatigue properties and toughness. It was found that the addition of the 0.5 and 1 wt% of multi-walled carbon nanotubes (MWCNTs) increase mode-I and mode-II fracture energy. The study [3] investigated the effect of adding multi-walled carbon nanotubes to the epoxy matrix of CFRP on impact resistance and impact damage progression. It has been shown that the absorbed impact and shear punch energies in the punch tests are higher for composite filled with MWCNTs by almost 21.3% which is connected with increased interlaminar fracture toughness, damping, and the alleviated crushing/buckling strength of the fibers.

The modified CFRPs laminates with 1.5 wt.% MWCNTs were subjected to low-velocity impact at three impact energy levels (8, 15 and 30 J) and directly compared with the unmodified laminates in the paper [4]. In terms of the CFRPs impact performance, the compressive strength of nanomodified composites was improved for all energy levels with the incorporation of 1.5 wt.% MWCNTs compared to the reference material.

It is important for proper investigation and comparison of properties to assess fiber-to-binder ratio. One of the robust methods is thermogravimetric analysis that is described in [5-7]. The present paper is devoted to the investigation of hybrid CFRPs modified by different carbon filler. Thermogravimetry is used to evaluate thermal properties of hybrid CFRPs and establish fiber content.

Content from this work may be used under the terms of the Creative Commons Attribution 3.0 licence. Any further distribution of this work must maintain attribution to the author(s) and the title of the work, journal citation and DOI.
Published under licence by IOP Publishing Ltd

2. Materials and testing technique

The specimens were carbon fiber reinforced polymers with additional carbon fillers. The lay-up was balanced, symmetric and pseudosotropic: $[0/90;+45/-45]_{4S}$. Hybrid CFRP components:

- CBX300 biaxial fabrics made of PAN carbon fiber (Mitsubishi Pyrofil TR50S 12K);
- R&G Epoxy L with GL2 hardener;
- single-wall carbon nanotubes (SWCNT) TUBALL™ by OcSiAl (denoted as CNT) or;
- MATRIX 201 by OcSiAl – commercially available 10%CNT+90%Alcohol solution modificator for epoxy resin (denotes as MTRX) or;
- milled carbon fibers (denoted as uCF).

The resulting thickness of the blank was ~4 mm and the specimens were cut from the blank using a milling machine to obtain nearly cube-shaped specimens: the thickness of the specimens was as-moulded; length and width were 4×4 mm. The content of the fillers was varied resulting in the formation of 10 specimens; the compositions tested are summarized in Table 1. The additive concentration is calculated for the total epoxy weight, not for the resulting CFRP. For each composition 3 blanks were prepared.

Specimens were tested using NETZSCH STA 409 PC/PG. Testing parameters are heating from ambient temperature to 873 K with a rate of 15 K/min and Ar flow of 50 ml/min.

Table 1. Overview of the tested compositions

Additive	**Amount, wt%**	**Designation**
Non-modified CFRP	0	NM CFRP
CNT	0.1	CNT-001
	0.2	CNT-002
	0.3	CNT-003
	0.5	CNT-005
uCF	0.5	uCF-005
	1	uCF-010
	2	uCF-020
	5	uCF-050
MTRX	2	MTRX-020

3. Experimental results and discussion

Figure 1 presents TG and DSC curves obtained for the NM CFRP specimen. The weight loss curve has 3 stages that can be identified. The first stage is from the start of the test till 320-350 °C and it is associated with the less than 1 wt% loss of the weight. Within this temperature range, the thermoset epoxy binder in the neutral Ar atmosphere is stable and no significant changes occur. Then after 330-360 °C a sharp drop associated with binder decomposition can be seen. This process of decomposition depends on the heating rate and gas flow and it can be seen that the weight loss slows down at 450-500 °C. The rate of heating used (15 K/min) is higher than 10 K/min used in investigations by different researchers thus it can be concluded that the process of decomposition does not finish completely at 600 °C. Thus the transition between the second and the third stage is nonlinear.

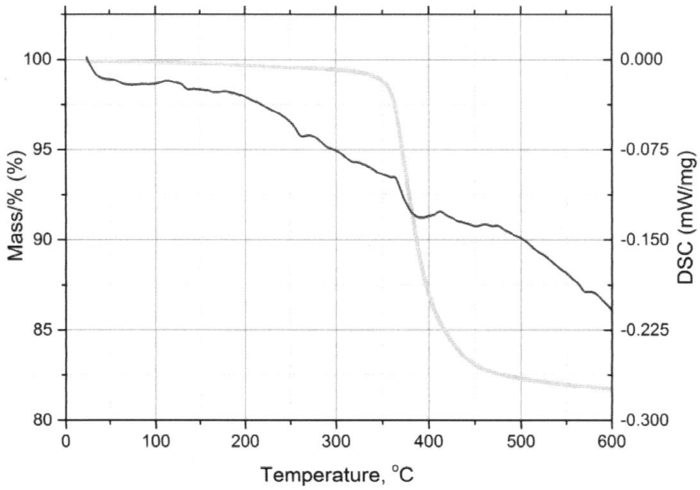

Figure 1. Typical TG (green) and DSC (blue) dependencies for non-modified CFRP specimen

Figure 2 shows the plot combining weight loss curves obtained for all specimens. Two insets describe detailed images of two areas: the transition from the first to the second stage and the second part of the stage 2. The former one describes the initial of the binder decomposition process and the latter one – its ending. It can be seen from Figure 2 that the first stage is characterized by the lower slope of all weight loss curves for CNT and MTRX specimens. Thus it can be concluded that the SWCNT modified epoxy binder has higher thermal stability. However, the transition from the first to the second stage (left inset in Figure 2) shows that non-modified CFRP and uCF specimens are grouped more on the right side, i.e. the process of decomposition of these specimens starts later than for CNT and MTRX specimens. It seems that the presence of SWCNT promotes the initial binder decomposition while the rates are nearly the same. The third stage is finished with a small slope of weight loss curve thus, as it was noted above, the binder decomposition is not completely finished.

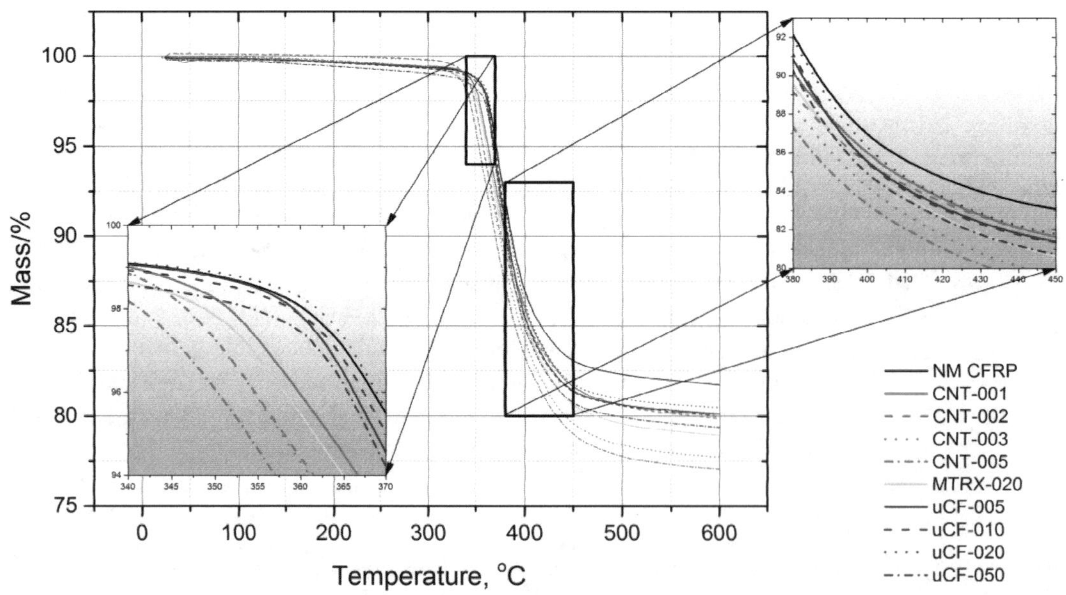

Figure 2. Weight loss curves obtained after TG for all investigated specimens

Final weight residue after TG testing (at 600 °C) was used to calculate the fiber content in fabricated and tested CFRPs. For these calculations we used datasheet density values of fiber end epoxy binder which are 1.82 g/cm³ and 1.15 g/cm³ correspondingly. The following equation was used for $CF_{content}$:

$$CF_{content} = \frac{w_R - 100\mu_A + w_R\mu_A}{\rho_{CF}/\rho_E(100 - w_R - 100\mu_A + w_R\mu_A) + w_R} \quad (1)$$

where w_R – weight residue after the TG test; μ_A – amount of additive; ρ_{CF} – density of carbon fibers; ρ_E – density of epoxy binder. The results of the $CF_{content}$ depending on the type and the amount of the filler are shown in Figure 3.

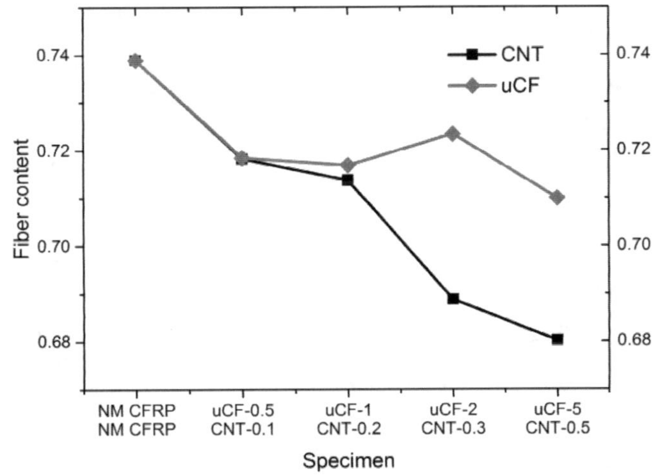

Figure 3. Fiber content vs type and amount of filler

4. Conclusion

Experimental TG/DSC testing of hybrid CFRPs modified by the addition of single-wall carbon nanotubes and milled carbon fibers was performed. It was shown that the weight loss (decomposition) curves have some differences. The rate of decomposition in the second stage does not differ and it is nearly constant but the initial of decomposition is shifted to the lower temperatures for SWCNT modified CFRPs. The residual weight of the specimens was used to assess the fiber content in the moulded CFRPs.

Acknowledgements
This work has been supported by the Russian Scientific Foundation, grant 19-79-10148

References

[1] Knoll J B, Riecken B T, Kosmann N, Chandrasekaran S, Schulte K and Fiedler B 2014 The effect of carbon nanoparticles on the fatigue performance of carbon fibre reinforced epoxy *Compos. Part A Appl. Sci. Manuf.* **67** 233–40

[2] Quan D, Urdániz J L and Ivanković A 2018 Enhancing mode-I and mode-II fracture toughness of epoxy and carbon fibre reinforced epoxy composites using multi-walled carbon nanotubes *Mater. Des.* **143** 81–92

[3] Tehrani M, Safdari M, Boroujeni A Y, Razavi Z, Case S W, Dahmen K, Garmestani H and Al-Haik M S 2013 Hybrid carbon fiber/carbon nanotube composites for structural damping applications *Nanotechnology* **24** 155704

[4] Dimoka P, Psarras S, Kostagiannakopoulou C and Kostopoulos V 2019 Assessing the Damage Tolerance of Out of Autoclave Manufactured Carbon Fibre Reinforced Polymers Modified with Multi-Walled Carbon Nanotubes *Materials (Basel).* **12** 1080

[5] Bücheler D, Kaiser A and Henning F 2016 Using Thermogravimetric Analysis to Determine

IOP Conf. Series: Materials Science and Engineering 1118 (2021) 012035 doi:10.1088/1757-899X/1118/1/012035

Carbon Fiber Weight Percentage of Fiber-Reinforced Plastics *Compos. Part B Eng.* **106** 218–23

[6] Moon C-R, Bang B-R, Choi W-J, Kang G-H and Park S-Y 2005 A technique for determining fiber content in FRP by thermogravimetric analyzer *Polym. Test.* **24** 376–80

[7] Yee R Y and Stephens T S 1996 A TGA technique for determining graphite fiber content in epoxy composites *Thermochim. Acta* **272** 191–9

AUTHOR INDEX

Abramov, Yuri ... 74
Afanevich, V V ... 42
Alushkin, T ... 128
Arkhipova, O V ... 166
Belozertseva, L Yu ... 1
Belyaev, A S ... 97, 113
Bogachkov, I M ... 80, 159
Bogdanovich, V I ... 20, 25
Bolotov, S V ... 14
Borisov, D S ... 65
Brylev, O A ... 113
Burkov, A V ... 183
Burkov, M V ... 188
Chernienkov, E M ... 7
Chudin, A A ... 65
Deryuzhkova, N E ... 90
Dolgikh, N N ... 166
Dovbysh, N S ... 119
Dronov, A I ... 70
Eremin, A V ... 183, 188
Eremin, S V ... 109
Filippov, V V ... 65
Frayshteter, V P ... 80
Giorbelidze, M G ... 20, 25
Gordynets, A S ... 124
Ivanov, E A ... 113
Kalashnikova, E K ... 132
Kayumov, Kh A ... 137
Khamitov, R N ... 80
Khasanov, S M ... 177
Khmelevsky, Y ... 103
Kholodilin, Ivan ... 74
Kidalov, S V ... 132
Kovalev, V Z ... 166
Kozlov, I V ... 65
Krasikov, P P ... 65
Krasikova, L S ... 65
Kravchenko, S M ... 141
Kruchynina, A S ... 55
Kulev, A V ... 109
Kuvandikov, O K ... 137
Kuvandikov, Sh J ... 137
Kuzin, E G ... 146
Lago, Luis A ... 74
Lomakin, D O ... 109
Lubyanoi, D A ... 146
Lustenkov, M E ... 30
Lustenkova, E S ... 36
Mamontov, G ... 103
Markidonov, A V ... 146
Meshcheriakov, M ... 128
Mitrofanov, A N ... 119
Moiseenko, A N ... 30
Negodin, A ... 151, 155
Nepomnyashiy, A S ... 124
Novikov, A N ... 109
Novikova, E A ... 119
Osipov, D S ... 166
Paramzin, A O ... 166
Parshin, S V ... 70
Pashkevich, V M ... 42
Polesskiy, O A ... 65
Popov, M ... 151, 155
Pudov, E Yu ... 146
Rabyko, M A ... 48
Raznoschikov, A S ... 60
Salakhitdinova, M K ... 137
Savinov, A V ... 65
Savosteenko, Nikita ... 74
Semenova, O S ... 146
Seryakov, V ... 103
Shakhmanov, V N ... 146
Shemenkov, V M ... 48
Shevtsova, A G ... 109
Shilko, V K ... 172
Slepchenko, I V ... 141, 172
Slepchenko, V A ... 141, 172
Solovyev, V A ... 90
Sumenkov, O Y ... 97
Svirepa, D M ... 55
Teterin, V V ... 90
Tkachenko, V A ... 166
Tkacheva, L ... 103
Vlasov, Yu ... 128
Voznesenskaya, A A ... 60
Vozniakovskii, A A ... 132
Voznyakovskii, A P ... 132
Yakubov, A A ... 177
Zhdanov, A V ... 60, 119

Institute of Physics
Dirac House, Temple Back
Bristol BS1 6BE UK

ISSN: 1757-8981
ISBN 978-1-7138-3404-5

CURRAN ASSOCIATES INC.
proceedings.com